16

Techniques of
Environmental Systems Analysis

TECHNIQUES OF ENVIRONMENTAL SYSTEMS ANALYSIS

R. H. PANTELL
Electrical Engineering Department
Stanford University
Stanford, California

A Wiley-Interscience Publication

JOHN WILEY & SONS

New York / London / Sydney / Toronto

Library of Congress Cataloging in Publication Data:

Pantell, Richard H   1927–
    Techniques of environmental systems analysis.

    "A Wiley-Interscience publication."
    Includes index.
    1. Environmental protection.   2.   System analysis.
I.   Title.
TD170.2.P36   363.6   76-98
ISBN 0-471-65791-3

□ Preface

During the past few years there has been an increasing interest in the management of environmental systems. As used in this text, "environmental systems" does not refer solely to physical elements, but also includes economic and social factors. This concern has resulted from a recognition of serious deteriorations in environmental quality (e.g., air pollution and urban congestion), as well as depletions of natural resources.

Thus far, however, systems analysis has had relatively minor impact on the resolution of environmental problems. The reasons for this disappointing result include:

- A lack of ability on the part of the analyst to communicate information in a meaningful manner to the decision maker.
- An overselling of the capabilities of systems analysis.
- Inadequate data.
- Uncertainties concerning cause and effect relationships between environmental variables, and difficulties related to ascertaining these relationships.
- A distrust, on the part of the public, of analytical approaches.
- An inability to express planning objectives in a manner that can be used to measure progress toward these objectives.

There are, however, a number of advantages to a systems approach, which is merely an attempt to find an orderly means for choosing between options. If properly applied, it can lead to a better understanding of the relationships between variables and of the value judgments that must be considered. Systems planning is not a panacea for environmental problems, but rather a decided improvement over a lack of planning.

In this text various aspects of systems analysis are presented. It is intended that the material be considered a challenge rather than a revelation. The challenge is to seek methods to improve the deficiencies associated with systems planning and thereby enhance the usefulness and impact of these procedures.

Chapter 1 is an overview of systems analysis. There is no unique 'best method' for the study of a complex environmental problem, but certain features should be present. A *formulation* step is required to proceed from what is usually a rather general definition of objectives to an understanding of the choices, a feeling for what can and cannot be controlled, and a qualitative picture of the components of the system and how they may interact. A *modeling* or *analysis* step is necessary to proceed from a qualitative to a quantitative description of the system. Finally, an *evaluation* should be performed to assign preferences to the various available alternatives. Chapter 2 is concerned with analysis by means of conservation equations. These equations describe the rate of flow of a commodity into and out of a system. The commodity may be money, a pollutant, biomass, or numbers of a species. Various subjects are considered, including birth and death of a population, transitions between different groups, water pollution, predator–prey, air pollution, and urban growth. The guiding principle in all of these problems is to equate the rate of increase of a variable to the rate of input minus the rate of output.

Chapter 3 is devoted to the economic aspects of environmental management. Financial concerns are usually a critical component of environmental systems, and therefore special emphasis is required in this area. The material includes a discussion of costs and benefits, a consideration of the failings of laissez-faire supply–demand pressures with regard to pollution and resource depletion, and a comparison of alternative methods of environmental management from the standpoint of the cost required to achieve a specified level of performance.

Various decision problems are considered in Chapter 4. Topics included are decision making under risk for a single attribute, how to choose an alternative under uncertainty, and techniques for handling systems with two or more incommensurate attributes.

I would like to express my gratitude to Dr. Morteza Kashef for reading early renditions of the manuscript and for contributing Appendix A, and to Mrs. Janet Dinkey for her secretarial assistance.

I am grateful to the Literary Executor of the late Sir Ronald A. Fisher, F. R. S., to Dr. Frank Yates, F. R. S., and to Longman Group Ltd., London, for permission to reprint the Table of Random Digits from their book, *Statistical Tables for Biological, Agricultural and Medical Research* (6th ed., 1974).

R. H. PANTELL

Stanford, California
October 1976.

□ Contents

Techniques of
Environmental Systems Analysis

Chapter One □ Systems Approach

World War II provided an impetus to the development of techniques for the analysis of complex systems that were primarily concerned with military matters. These systems included consideration of economic, technological, political, and human factors. An excellent discussion of many of the procedures and difficulties encountered in military systems is presented in E. S. Quade and W. I. Boucher (eds.), *Systems Analysis and Policy Planning, Applications in Defense,* American Elsevier, New York, 1968. Similar methods have been applied to business management, drawing material from the subject headings of operations research, systems analysis, decision theory, cost benefit, cost effectiveness, and a variety of other titles. The reader may be interested in some of the definitions of and distinctions between operations research,[1] systems analysis,[2,3] decision theory,[4-6] and cost–effectiveness studies.[7]

Features of Systems Analysis

These approaches attempt to apply procedures to the selection of options or alternatives that are more explicit, and hopefully more consistent than procedures that have been used in the past. The elements of a systems approach to problem solving are as follows: (a) a specification of possible alternatives, (b) a determination of the consequences of following each alternative, and (c) the use of objective statements or a value system to indicate whether one outcome might be preferable to another. It is not to be implied that the procedures are unique or without uncertainty. On the contrary, since planning involves projections in time there is always a minimum uncertainty as to the conditions that will prevail in the future.

A rather broad definition of systems analysis is provided by Quade and Boucher, "A systematic approach to helping a decisionmaker choose a course of action by investigating his full problem, searching out objectives and alternatives, and comparing them in the light of their consequences, using an appropriate framework—in so far as possible analytic—to bring expert judgment and intuition to bear on the problem." It is difficult to define precisely an activity that covers such diverse subject matter, and it is probably not worth the effort.

Some desirable features of an appropriate application of systems analysis are:

- The steps in reaching a decision are specified so that the procedures may be assessed, improved upon, or reproduced.
- The value judgments applied to the decision process are explicitly indicated.
- The effort to relate effect to cause frequently results in an improved understanding of the problem.

The relevant question concerning the use of the procedures described in this chapter is whether or not it is an improvement over existing or alternative procedures. Perhaps the most useful application of systems analysis in the decision process is as an adjunct to rather than a replacement for the judgment of a planner. A planner or decision maker may have accumulated experience that enables him to perceive overall strategy considerations that would be difficult to express in a formal manner. For example, we can build computers that are excellent checker players, outstanding but not first-quality chess players, and rather poor go competitors. As the strategy of the game increases in complexity there is more difficulty in the modeling. Of course the capacity of computers is increasing faster than the capacity of the human mind and eventually we may achieve a crossover. Even at the present time, however, an important role for systems studies is to provide information to a decision maker describing the effects of changing parameters of the system, changing one's value structure, or substituting one alternative for another.

Analysis Applied to Environmental and Social Systems

In recent years, the systems approach has been extended to include environmental and social problems such as pollution,[8,9] transportation,[10] land-use management,[11] education,[12] health-care delivery,[13,14] and criminal justice.[15] As the procedures have been extended into new areas, so the criticisms have mounted.[16] Some of these criticisms are valid. A good deal of modeling has been based on unproven structural and behavioral assumptions. (Structural assumptions refer to the manner in which one element of the system is related to another, and behavioral assumptions concern human response to imposed conditions or changes.) The fact that social and environmental systems studies have had very little effect on the world around us is at least partially a consequence of inadequacies of the modeler and his analysis. Not the least important obstacle is the inability of the systems analyst to communicate his information in meaningful terms, or to recognize the social, political, and economic constraints imposed on a decision maker. This does not necessarily mean, however, that the approach should be abandoned, but rather that it should be improved.

1.1
SOME CRITICISMS AND CONCERNS ABOUT SYSTEMS STUDIES

Reduction to a Single Number

It is not common to initiate a discussion on a defensive note, but prejudices about the employment of systems analysis for environmental and social problems are sufficiently prevalent to warrant a response. One concern is that an analyst may attempt to reduce many incommensurate factors to a single number, so as to enable selection of one alternative from many. How, for example, may a problem involving esthetics, cost, ecosystems, and housing be reduced to a single number by which to compare one's choices? At first this appears to be an inappropriate, if not impossible task but, on the contrary, it is unavoidable.

A description of a system in terms of many variables, either given qualitatively or measured in different units, is a vector description. The information about the problem cannot be presented as a single measurement. Each component of the vector will alter as one alternative plan is substituted for another, and an alternative may be described in terms of the vector it produces. However, when an alternative is selected the title "most desirable" involves only a single factor, desirability, and therefore a scalar description applies. This means that an operation must be performed on the vector to obtain a scalar. At the present time this transformation is not usually stated explicity, but one function of systems analysis is to bring the operation into the open where it may be investigated for reasonableness and consistency.

Quantification

Another criticism is that the systems approach tries to quantify all variables, whereas some aspects are not quantifiable. A similar complaint is that those variables most readily quantified, such as time and cost, will receive disproportionate emphasis. Quantifiability is not equivalent to importance. This is an important problem without a simple answer. A systems analyst can avoid consideration of those variables that are difficult to quantify, such as esthetics, and focus only on those measurable on some convenient scale. The unconsidered aspects may be left to the decisionmaker to synthesize with the other data. The latter may, for example, impose implicit constraints on esthetic impacts based on his own preferences or his perception of community perferences. Alternatively, the analyst may incorporate esthetics in his investigation by: (a) seeking measurable indicators for esthetics (e.g., how many acres of open space are to remain), (b) having the decisionmaker or other interested party perform a preference ranking or worth estimate for possible alterations in the appearance of the environment. Obviously there is no unique best procedure for the inclusion of certain types of difficult-to-quantify factors.

Uncertainty

An additional concern regarding environmental and social systems analysis is the degree of uncertainty that exists. How can one plan in any formal manner when there are uncertainties about the future, when data are lacking or uncertain, and when the structure of the system is ambiguous? Undoubtedly the existence of uncertainties complicates the planning process, but again the relevant question is whether a lack of systematic planning is preferable when there is uncertainty? Rather we should seek reasonable techniques for handling uncertainty, such as investigating possible outcomes over the uncertainty range of important parameters, or perhaps selecting the alternative that has the best performance when the uncertain variables are at their worst possible values.

This does not exhaust the list of objections, and those who wish to pursue the subject further can find ample discussions in the literature.[16]

1.2
METHODOLOGY FOR ANALYSIS

With the growing recognition of a need for planning, management models have developed in a variety of fields. In this section we shall review planning procedures devised for a variety of different problems, and from this study a proposed general approach to planning will be presented.

Approaches to Analysis

Figure 1.1 is an illustration of a management cycle for an educational program.[7] The arrows show a flow of information and the boxes indicate

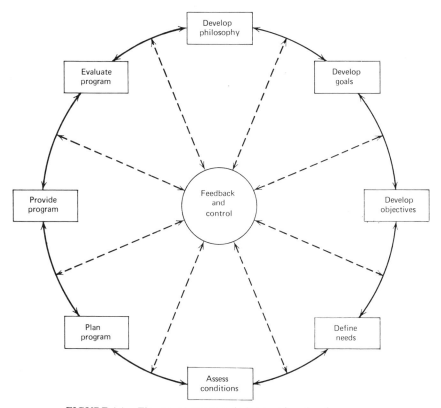

FIGURE 1.1. The management cycle for an educational program.

operations. If the management process is begun at the top of the diagram and we proceed in a clockwise direction, the first three boxes, *philosophy*, *goals*, and *objectives*, will refer to statements of the purpose of the effort. Each box becomes more specific, with *philosophy* referring to very general concepts and *objectives* written in terms of factors that can be verified or measured. The next three boxes, *needs, conditions*, and *program plan*, are the stages of the planning operation. *Needs* refer to expressions of the gaps between the status quo and the desired situation. *Conditions* are descriptions of the relationships between the variables that the planner can control and those that are to be controlled. *Program plan* is a consideration of the alternatives that might be employed to achieve the objectives and a determination of the necessary resources. In the final two boxes, the program is introduced and an evaluation is made of the effectiveness of the chosen course of action. The purpose of the feedback–control circle in the center is to ascertain that the planning done at any given stage is consistent with what has gone before, and to indicate modifications when inconsistencies develop.

As an illustration of the use of this cycle, suppose our *philosophy* statement is that all children in our school district should have equal educational opportunities. As a *goal*, every child should obtain equivalent instruction in science. An *objective* is to provide comparable facilities for science instruction for all children in the state. A *need* is identified from an inventory of science instructional facilities, whereupon it is found that some school districts have extensive equipment and others have none. *Conditions* for equalizing facilities are determined in terms of space and financial requirements. Alternative methods for meeting the need are considered in the *program-planning* stage. These alternatives include science learning centers, which will bring students from several schools to a central science facility, or upgrading the facilities at each school where a need has been identified. Costs and educational and social benefits of each alternative are considered and one plan is accepted. The program is then implemented and an evaluation made to determine whether each child has access to comparable facilities for science instruction.

A similar planning schedule, shown in Fig. 1.2, was developed for environmental health problems by Prof. A. L. Delbecq at the University of Wisconsin. Step 1 in Fig. 1.2 is approximately equivalent to the first three boxes in Fig. 1.1. *philosophy, goals,* and *objectives.* Step 2 corresponds to the next two boxes. Step 3, *final priority determination,* was not explicitly

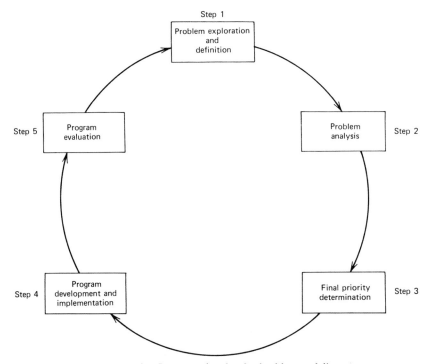

FIGURE 1.2.　Program planning for health-care delivery.

included in Fig. 1.1. If a multiplicity of needs has been identified, with limited resources it may be necessary to assign a priority to the problems. Step 4 in Fig. 1.2 is approximately equivalent to the *plan program* and *provide program* boxes in Fig. 1.1, and Step 5 is the *program-evaluation* box in Fig. 1.1. Figure 1.2 does not show the *feedback–control* mechanism in Fig. 1.1, but it would be unlikely that the steps in Fig. 1.1 could be performed in a consistent manner without reconsideration of previous conclusions.

A model proposed for urban renewal planning[11] is illustrated in Fig. 1.3. The operations shown in this figure are essentially the same as those shown in Fig. 1.1, except that the process terminates with the recommendation of a program and does not include the implementation and evaluation of the program. The inputs to the model consist of supply–demand information for residential requirements, as well as alternative programs under which the demands are satisfied. Additional inputs are the behavior variables,

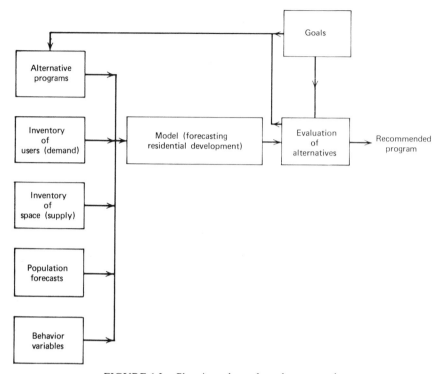

FIGURE 1.3.   Planning scheme for urban renewal.

which specify the preferences of the people for various locations and living conditions. Land allocation for residential development is obtained from the *model*, utilizing the inputs and hypothesized economic principles relating allocation to supply and demand. The goals of the community or planning commission are used to formulate the alternative programs and to evaluate the output from the model for each alternative considered. These evaluations may be used to modify the alternatives if no acceptable alternative had been presented, and the cycle then repeats. The output from the *evaluation* effort is a recommended program.

A comparison between Figs. 1.3 and 1.1 shows that the essential steps in the planning effort are the same, regardless of the particular problem considered or the specific format for the development of a plan. The *goals* box in Fig. 1.3 corresponds to the *philosophy*, *goals*, and *objectives* boxes in Fig. 1.1. *Model* inputs in Fig. 1.3, consisting of *inventory of users and space*, *population forecasts*, and *behavior variables*, would be considered in the *define needs* box of Fig. 1.1. The *model*, relating input to output variables, is approximately equivalent to the *assess conditions* box in Fig. 1.1. *Alternative Programs* and *Evaluation of Alternatives*, as shown in Fig. 1.3, correspond to the *plan-program* step in Fig. 1.1.

As a final example, the methodology used to analyze alternative regional transportation systems is depicted in Fig. 1.4.[10] The initial step is a description of the environment in which the transportation network will operate, including possible technological and social changes that may occur and various future population distributions over the planning interval. (In this particular case the planning period is 1970–2000.) Basic transportation networks and services are also specified. The next box in Fig. 1.4 indicates that the base mix of transportation modes (auto, bus, rail,

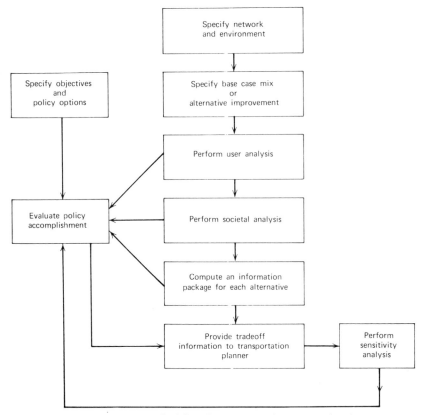

FIGURE 1.4.   A methodology for transportation-system planning.

and air) is given (i.e., the network that would develop from 1970 to 2000 without decisions to change the system), as well as alternative improvements to be considered.

An analysis is performed to determine the impacts of each alternative network on the users and on the rest of society. These impacts include reference to such items as system reliability, meeting demand for passenger and freight transportation, reducing congestion, and minimizing detrimental environmental effects. An information package is then prepared, describing the effects of each alternative.

The selection of an alternative requires a comparison between the objectives of the groups affected by the network and the impacts resulting from following an alternative. Therefore, evaluation has inputs that consist of impact information and objective statements. Tradeoffs, which indicate the consequences of placing a different relative emphasis on the various objectives, are described to the transportation planner. The sensitivity analysis is a study of the effects of varying the parameter values for those variables for which a high degree of uncertainty exists (e.g., population distribution, technological advances, and demands for transportation). It may be preferable to choose an alternative having satisfactory performance over a broad range of variable values rather than an alternative having outstanding performance over a narrow range of parameters.

We see that there is not a unique way to organize the planning process, but we do note that there are common elements in all of these approaches. A procedure that incorporates the steps presented in Figs. 1.1–1.4 is shown in Fig. 1.5. The *formulation* step consists of an identification of the purposes of the effort, supply–demand considerations, and a listing of alternative solutions. *Modeling or analysis* relates the input variables of the system, such as population projections, economic growth, or transportation network, to output variables, such as the effects on the natural and social environments. Each alternative strategy is *evaluated*, based upon a comparison between the outputs from the model and the goals of the planning effort.

In the following sections of this chapter we shall indicate some of the approaches to environmental and social systems planning and some of the difficulties that will be encountered. We will not consider the program implementation and program evaluation stages included in Figs. 1.1 and 1.2, but rather will limit our attention, as in Figs. 1.3 and 1.4, to *formulation, modeling*, and *evaluation*.

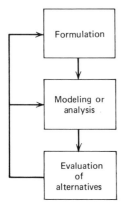

FIGURE 1.5. An approach to planning.

1.3
FORMULATION

The investigation often begins with a vague statement of goals, and it is the purpose of the formulation operation to identify the critical components of the system and to seek a qualitative understanding of the interactions between the components. Table 1.1 lists the steps generally taken in the formulation, although the sequence of items does not necessarily indicate the order in which these steps are performed. If, for example, a problem is primarily technical in nature, it might be appropriate to first identify the input and output variables, whereas a problem with strong social implications might be attacked with the sequence listed in Table 1.1.

Table 1.1
Formulation

1.	Determine the interested parties.
2.	Specify goals and criteria for evaluation.
3.	Determine the state of the art and what additional information will be required.
4.	Identify input, output, and control variables.
5.	Enumerate alternatives.
6.	Perform a qualitative structuring and modeling of the system.
7.	Estimate future conditions, including variable projections and technological changes.

Interested Parties

An initial step is to indicate the interested parties. Figure 1.6 illustrates the groups concerned about transportation systems: (a) users, (b) operators, (c) other members of society, and (d) implementing agencies. (The transportation example described in this chapter is based on material presented in reference 10.) Each of these parties may be subdivided further, as follows: (a) users are grouped into passengers and freight, (b) operators are regulated or nonregulated suppliers or governmental, (c) other societal groups are categorized by income, and (d) implementing agencies are categorized as planning, financing, regulatory, and miscellaneous. As shown in Fig. 1.6, further subdivision may be desired so that, for example, passengers are traveling for either business, pleasure, or commutation.

There is no simple rule for determining the extent to which the various groups should be subdivided. The members of each group are considered to be identical in their interests and properties, and this is an obvious approximation since no two individuals are identical. When groups are disaggregated further, the purpose is to account for nonuniform properties and thereby increase the accuracy of the analysis. However, as the number of groups increases, so does the complexity of the problem. The level of disaggregation is usually determined by classifying groups on the basis of those properties believed to influence the output variables. For example, if demand were to be satisfied by a transportation network, and the spatial and temporal dependencies of demand varied with the pleasure or business nature of the passengers' travel, it would be appropriate to classify the users according to whether the trip were for pleasure or business.

Goals

Item 2 in Table 1.1 indicates that the goals of the program are to be specified and, in general, each of the interested parties may have different goals. For example, considering the passenger and societal groups in a transportation system, the goals may be:

Passenger	Society
1. Minimize travel time.	1. Minimize detrimental effects on human physiology (e.g., pollution, accidents.)
2. Minimize travel cost.	
3. Minimize anxieties concerning safety and security.	2. Minimize inconveniences associated with right-of-way (e.g., the

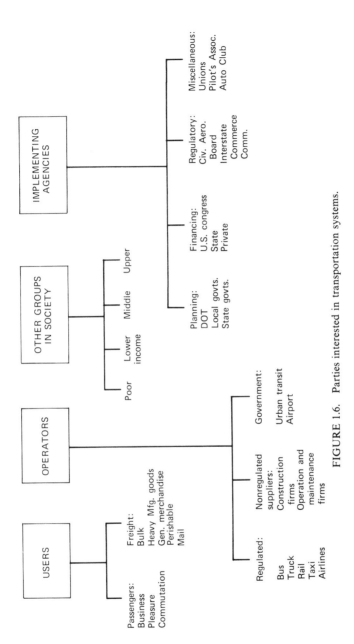

FIGURE 1.6. Parties interested in transportation systems.

Passenger	Society
4. Maximize convenience.	displacement of property owners, obstruction).
5. Maximize comfort.	3. Maximize economic benefits (e.g., provide employment, distribute industry).
	4. Improve design and form of metropolitan areas (e.g., relieve city-center crowding).
	5. Satisfy sociological needs, (e.g., provide more leisure time and access to schools).
	6. Minimize further reduction of resources (e.g., minimize changes in ecosystems and effectively utilize energy).

Interested party groups are not necessarily mutually exclusive. A passenger is also a member of society, and therefore a single individual may have goals common to both groups.

An important reason for stating goals is to be able to define attributes or indices associated with the goals by which relative benefits of different alternatives can be measured. This is accomplished by continuing to subdivide the goals until attributes are obtained that can be verified either by a direct scale measurement (e.g., dollars for cost) or by a worth estimate (e.g., a rating from 0 to 1 for the esthetics of a vehicle). Table 1.2 illustrates the breakdown of passenger's goals for travel time and comfort into attributes.[10]

Similar attribute assignments may be made for the other goals. The measured values for the attributes change as one alternative is substituted for another, and it is on the basis of these values that the benefits, losses, and tradeoffs can be determined for each alternative. The goal statements and associated attributes are not static, for insights gained from later stages of the planning will often suggest modifications of the original set of goals and objectives.

The goals for different interested parties may conflict, or there may be conflicts between goals for a single group. An action intended to meet passenger demands in a transportation system might generate undesirable

Table 1.2

ASSIGNMENT OF VERIFIABLE ATTRIBUTES TO THE GOALS

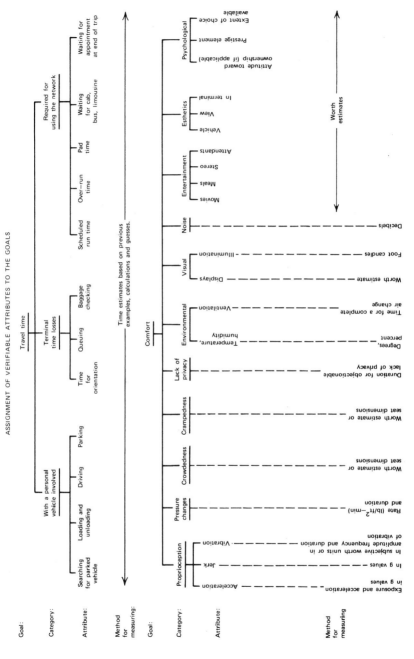

ecological consequences. Conflict may be handled in a number of ways. The systems analyst can vary the relative emphasis or weight placed on the different goals and determine the preferable alternative for each set of weights. This procedure informs the interested parties of which plan to choose for any given priority assignment to the goals and leaves the final selection to the responsible individuals.

Another approach is to produce a social welfare function (i.e., a preference ranking of alternatives or a goal weighting scheme for society as a unit) based on the preferences of individuals or different groups. Considerable thought has been devoted to the determination of a social welfare function, but inadequacies exist in all proposals.[17] Majority rule is often the basis for settling conflicts in the United States, and it has the advantages that it always leads to a unique group decision and gives the same authority to every individual. In some instances it might be desirable to assign a weighting factor to the vote of an individual, based on the extent to which that person is affected by the decision. A problem regarding the location of Indian reservations in the United States affects the majority in a minor way, but a small minority in a significant way. There should be some recognition of this difference.

State of the Art

As indicated by the third item in Table 1.1, it is necessary to become familiar with existing conditions, such as: (a) current facilities and apparatus relevant to the planning effort, (b) data that may be used to estimate the consequences of different actions, (c) previous proposals or current plans for achieving the goals, (d) the present transportation network, (e) existing plans for extending urban transit, (f) projections, if any, for population distributions to the year 2000, (g) relevant publications, and (h) the local experts. Gaps in the information may become apparent, and these should be noted to indicate areas where additional investigation may be necessary.

The Variables

The fourth item in Table 1.1 refers to the identification of input, output, and control variables. Input variables are those factors that are imposed upon the system. Some of these variables, termed exogenous or independent, are assumed to remain unaltered by the introduction of the system. For a transportation-network analysis, this would include distances between cities, population projections, technological factors, and climatic conditions. Output variables, also known as dependent, endogenous, or state variables, describe the performance properties of the system. The output variables of interest to a passenger are the time, cost, convenience, comfort, and anxieties of travel. These variables are specified by the goals and concerns of the interested parties and are measured by their corresponding attributes (see Table 1.2). A given variable might be either exogeneous or endogeneous, depending on the problem. In some instances a population projection may be insensitive to changes in a region and therefore would be exogenous. On the other hand, if influences on population growth were of concern, this would be an endogenous variable.

The values assumed by the output variables are determined by the values for the input variables and the particular system or network that is introduced. Certain factors in the design of the system may be controlled by the analyst and planner, and these are the control or decision variables. In a transportation network, the decision variables might be the locations of the links between cities, the mix of various modes of travel (air, bus, rail, and auto), the handling capacity of each mode, and the times at which new links or modes are introduced. A primary function of the analysis is to arrive at the most desirable values for these decision variables.

Alternatives

Each setting of the decision variables corresponds to a distinct alternative course of action (see item 5 in Table 1.1). If none of the proposed alternatives initially provide satisfactory results, additional alternatives would then be considered. The alternatives analyzed for the transportation network were:

1. Evolutionary growth of current service (the base case).
2. Addition of tracked air-cushion vehicle (TACV) service on high-density links and short-takeoff-and-landing (STOL) aircraft on low-density links, at an additional cost of $5,000,000,000.
3. Introduction of electric passenger cars and buses that operate on city streets at conventional speeds and at higher speeds between cities, at an additional cost of $5,000,000,000.

For each alternative, values are determined for the output variables (i.e., numbers are assigned to the attributes, such as those listed in Table 1.2).

An interesting example of alternative selection follows from an experiment in communication between two cities that was performed in a Balkan country during the mid 1960s. A telephone call was placed, a man set out in an automobile, another person was dispatched on a bicycle, a letter was mailed, four carrier pigeons were released, and a telegram was sent. Two carrier pigeons were the first to arrive with the message, but the other two never appeared.

If only one system may be developed, then the appropriate procedure is first to optimize each alternative with respect to its own decision variables. The bicyclist will be sent on the shortest route, an expert driver will be chosen for the car, the pigeons will be sturdy, and so on. Then the alternatives are ranked according to some prescribed criterion. If speed is the criterion, the carrier pigeons might be the best choice, but for reliability it might be the man on the bicycle. The criterion could consist of a combination of speed and performance, with some weighting factor assigned to each variable.

Structuring

It is often useful to perform a qualitative structuring and modeling of the system (see item 6 in Table 1.1). Structuring refers to a subdivision of the problem into smaller components, and modeling is an attempt to understand the interactions between the components. The reasons for structuring and modeling are as follows:

1. It is usually easier to analyze a subdivision of the problem rather than tackling the entire problem as a single entity.
2. The assignment of tasks to individuals in the group is facilitated once the system has been structured.
3. Qualitative relationships, such as who influences what, may be identified.
4. It is generally simpler to recognize what additional information is required to obtain a more complete understanding of the system once the problem has been modeled.
5. The number of variables for the entire problem might exceed the available computer capacity or be expensive to run, so that subdivision may be necessary from this standpoint.

Structuring and modeling are neither unique nor static in that there are a variety of ways to perform these operations, and it is most unlikely that any particular model would remain unaltered throughout the analysis. Indeed, one generally progresses from a rather coarse to more refined models.

One method for structuring a problem is to consider distinct time intervals with decisions and evaluations made for each separate interval. There are a number of reasons why time subdivisions are useful, Time-dependent variables may be considered constant if the percentage change during a period is sufficiently small. Another reason for using time subdivisions might exist if there were discrete changes in the variables at certain specified times. For example, if the financial resources of a project were augmented every 3 years it might be appropriate to consider 3-year intervals. Finally, separation into time periods could allow the planner to make decisions sequentially rather than simultaneously. In this manner the number of variables to be considered in any time interval is less than the number of variables involved if the problem is attacked as a single unit.

Alternatively, one may divide the problem on the basis of disciplines. From a personnel standpoint a disciplinary division is convenient because the appropriate distribution of tasks and authority to various individuals is easier to identify. However if strong interdisciplinary interaction is desired in the planning process, a disciplinary structure may not be appropriate since this will tend to reduce contact between individuals from different disciplines.

A problem may be structured on the basis of interested parties. A transportation study, for example, can be investigated from the standpoint of the user, the operators, society, and implementing agencies.

Another way to structure a problem is on a geographical basis, where different localities are treated separately. A geographical division would be useful if introducing a change in one area did not cause significant changes in other areas.

Various subdivisions may be further divided. For example, the development over the first time period may be structured on a geographical basis. The more the overall problem is subdivided, the smaller will be the number of decisions to be made in each subunit and therefore, the simpler will be the analysis. However each subunit interacts with all other subunits to varying degrees so that the behavior of the whole is not merely the sum of the behaviors of each unit considered separately. The greater the number of subunits, the more complex is the determination of the operation of the entire system given the operation of the subsystems. It is generally useful to divide the problem so that the interaction between the units is relatively small or can be easily specified. If further subdivision leads to a marked increase in the complexity of the interactions between units, it is not appropriate to divide further. In every problem a compromise must be made between smaller subunits, which are easier to handle, and the complexity of resynthesizing the whole from its parts.

One aspect of structuring is to define the boundaries of the problem. The study may be limited to a fixed area, a specified time period, a restricted set of interested parties, and a limited number of input and output variables. Hopefully one can define a set of boundaries wherein the problem is reasonably self-contained, but regardless where the boundaries are drawn there is the inevitable question, "Why wasn't such-and-such included in the study?" The setting of boundaries involves the compromise between having a tractable problem and including the critical factors.

Structuring may be on the basis of sources and flows of commodities. The commodity may be money, materials, people, air pollutants, and so on. Figure 2.9, for example, is an illustration of structuring on the basis of the sources and flows of materials.

Another alternative is to structure by consideration of the operations to be performed and the flows of informations. This type of breakdown is shown in Fig. 1.7 for the study of a transportation network. The population, an exogenous variable, produces a demand for transportation. With this demand, each alternative modal mix is applied to the design of a network. The requirements of implementing agencies and operators impose constraints on the network. When the network design has been established, the impacts of the system on the user and society can be determined. Figure 1.7 does not include a consideration of how these impacts might influence the population distribution. This feedback loop, shown as the dashed line in Fig. 1.7, would be important if one wanted to consider, for example, the redistribution of an urban population with the introduction of an urban mass-transit system.

Projections

The final item in Table 1.1, item 7, refers to the fact that the systems analyst must make estimates of future conditions. These estimates may include projections of population and economic growth, possible technological advancements, or changing attitudes. Obviously there will be uncertainty in these projections, and one method for handling this is to perform the analysis with different input values for the uncertain variables. For example, a transportation study for the Northeastern States assumed two population distributions in the year 2000. One distribution was based on an extrapolation of existing trends, and the other was an assumed population distribution that would result from policies intended to stimulate

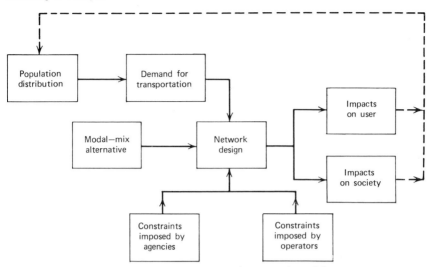

FIGURE 1.7.   A transportation-network model.

growth in the western portion of the region. Both these projections were used with each modal mix alternative to design and analyze the transportation system.

Not all of the steps listed in Table 1.1 will be applicable to every problem and, as previously mentioned, the order of tasks is not unique. These distinctions are not important. It is important, however, that for every analysis there be a formulation operation that proceeds from rather general policy statements to an identification of critical variables and a qualitative understanding of interrelationships between variables. Formulation and reformulation generally continue throughout the entire effort. As information is obtained from later stages, there may be a redefinition of goals, a change in structuring, or a consideration of additional alternatives.

1.4
MODELING OR ANALYSIS

In the formulation stage a qualitative understanding of the problem is sought, and in the modeling stage the understanding is quantified. This is accomplished by obtaining functional relationships between the variables so that for a given alternative and set of input values, the outputs are specified. There are a variety of forms for these relationships, such as equations, graphs, or tabulated numbers. One important function of modeling is to assist the systems analyst and decision maker to obtain a clearer understanding of the interactions between the different elements of the system.

An initial step in the analysis is to gather information that may be used to specify relationships between the variables. Whenever possible, these relationships should be checked for accuracy and, finally, any uncertainty in the parameters should be identified. Table 1.3 summarizes these steps in the modeling stage, and in the remaining portion of this section we shall consider the various aspects of modeling in the sequence listed in Table 1.3.

Table 1.3
Modeling

1. Gather information for relating the variables.

2. Specify the relationships between variables.

3. If possible, check the relationships.

4. Indicate assumptions, and identify uncertainties.

A first step in modeling is to gather information that can be used to relate the variables. Information may be obtained from data presented in the literature, from opinions given by experts, or from experiments performed by or for the systems analyst.

Use of Existing Data

There are several pitfalls associated with the use of existing data.[18] One difficulty is that usually the data are not collected for the specific problem concerning the analyst, and therefore, may not be applicable. As an example, suppose it is necessary to know the dependence of the spelling ability of primary school children on their weight. This is an incomplete statement of the problem, because spelling ability will also vary with age, economic status, education level of the parents, and so on. In the determination of how spelling ability varies with weight it must be indicated whether the other factors are held constant or allowed to vary. If a statistician collects data in a school yard from children in the first to sixth grades, he would find a correlation between weight and spelling ability. The primary reason for this is that both spelling ability and weight increase with age. It would not be correct to assume that augmenting the daily caloric content of a child's diet would improve his or her spelling ability. The data were taken with age allowed to vary and the experiment is performed with age held constant. This does not mean that the data are invalid, but rather that they be applied only for the circumstances in which they were taken. If one were to wander blindfolded through a school yard and was only given the weight of a child, it would be best to choose a heavy contestant for a spelling bee.

The previous example may be expressed in mathematical terms. Let us assume that spelling ability (as measured by an examination) S is a function of the age a of the child, and that weight w, age, and food consumption f are functionally related. The first experiment, in which a measurement is made of the relationship between S and w with a allowed to vary and f held constant, may be expressed as

$$\left(\frac{\partial S}{\partial w}\right)_f = \frac{dS}{da}\left(\frac{\partial a}{\partial w}\right)_f,$$

where the subscript f indicates that food consumption does not vary. It is seen from the above equation that spelling ability changes with weight because of the relationship between age and weight. The second experiment, with age constant and food consumption changed, is expressed as

$$\left(\frac{\partial S}{\partial w}\right)_a = \frac{dS}{da}\left(\frac{\partial a}{\partial w}\right)_a = 0.$$

Since age does not change with weight when age is held constant, spelling ability would not vary with weight when age is fixed and food consumption changes.

A relevant factor in all data gathering is the method for taking the data. There is evidence, for example, that reported increases in crime rates are at least partially a consequence of improved methods for reporting and recording crimes. Comparisons between statistics taken at different times or places or by different people should include a consideration of variations in the techniques for data gathering.

A third problem in the use of statistical data concerns the accuracy of the information. Every experiment perturbs the environment and may lead to erroneous conclusions regarding the state of the environment in the absence of the experiment. In addition, the measurement technique will introduce uncertainty in the data since it is generally impossible to include the entire population in a sample or measure any variable with complete accuracy. Finally, a time lapse between the accumulation of data and its application leads to uncertainty because of the time dependence of variables.

Information from Experts

In some instances it will be found that there are gaps in the data, and experts may be called upon to relate output to input variables. In addition, experts may be asked to:

- Evaluate data in terms of reliability and applicability.
- Identify alternatives and control variables.
- Indicate the attributes that may be used to measure the concerns and goals of the interested parties.
- Suggest the relative importance of goals and attributes.

For example, an art critic might be requested to designate indices (i.e., attributes) for measuring the level of creativity among contemporary artists. Perhaps he will specify that variety and quantity must be considered. He may also suggest procedures for measuring variety and quantity, and give his opinion of the relative importance of these two variables.

Some aspects of an expert of interest to the systems analyst are his abilities to apply his knowledge to the particular problem, predict events with reasonable accuracy, communicate the necessary information to a nonexpert, and be relatively unbiased. It is difficult to judge the ability of an expert. Such factors as his years of professional service, publications and academic rank must be considered. (On the one occasion that I had to judge the competence of a surgeon, I considered his ability to communicate the problem to me, sought the opinions of other doctors I knew personally as to the ability of the man in question, determined whether he had been involved in malpractice suits, and checked the publication dates of the most recent medical journals I could find in his office.)

The estimates of an expert will vary with his bias. For example, one would tend to discount the opinion of a doctor with regard to the hazards of smoking if he is employed by a tobacco company. Similarly, one would minimize the opinion of a transportation expert whose employer is contemplating an action that might result in traffic congestion. It is not that these people are dishonest, but rather that one's a priori beliefs are influenced by the extent of one's personal involvement. The a priori belief is the opinion of the individual or expert before any additional evidence is presented. The objective evidence generally allows for a range of interpretations, and a priori beliefs emphasize that portion of the evidence closest to the a priori condition. If a person is convinced that Germans are humorless, upon visiting Germany he will be most sensitive to dour expressions. If a person believes the Germans are easy going, he will be most sensitive to the smiling faces.

When there is disagreement between experts in the same field, a method known as the Delphi technique has been used to arrive at a consensus.[19,20] One procedure is for each expert to respond to a questionnaire, giving his opinions and the reasons for his opinions concerning the questions that are posed. This information is then distributed, in summary form, to all the experts, and they are requested to reconsider their positions in the light of this new information. It is hoped that this exchange of opinions will, in a few iterations, lead to a consensus (or a relatively small standard deviation of opinions). The Delphi approach avoids a direct confrontation and provides an anonymity of opinions, so that a particularly articulate or renowned expert does not exert undue influence.

Experts and nonexpert interested parties may interact directly with the model by means of operational gaming.[21] In this approach an expert may be presented with a status quo situation and a set of goals. He then makes a decision intended to bring him toward his goal, and this action is fed into a model of the system. Any alterations in the status quo, as predicted by the model, are then fed back to the expert and he may modify his action or proceed in any manner he considers appropriate. This process continues until the player is satisfied that the goals have been met, or a stalemate occurs, or a fixed playing time elapses.

Any number of experts and interested parties can play the same game. Gaming obviates the necessity to incorporate assumed behavorial patterns into the model, for human reactions and decisions are introduced by the players. In addition, gaming provides the players with insights concerning the consequences of their actions.

Of course, gaming is an approximation to an actual system and, therefore, is susceptible to errors. For example, the participants are not under the same pressures in a game as in real life, for they realize that a mistake in the game is not as costly as a real life mistake. Therefore they may be more adventurous than usual. Also, the model is an idealization of the system so that the information given the player may contain inaccuracies.

Data Gathering

In addition to the use of data and expert opinion, information may be ·
obtained from direct experimentation. The advantage of an experiment is
that it can be designed to fit the particular needs of the systems problem.
As mentioned previously, the use of another person's data may lead to
ambiguities concerning which factors are allowed to vary, whereas the
experimenter may have control over these factors. Experiments involve an
expenditure of time and money, and this cost must be balanced against the
benefit. If, for example, it is found that a wide range of values for a given
variable does not alter the decision, it would not be necessary to determine
this variable more precisely. Alternatively, a rather simple experiment may
sometimes be designed to pinpoint the value of a particularly critical
variable or parameter.

A common form of experiment, often used to determine value judgments, is the questionnaire. Some problems associated with questionnaires
are:

• If less than 100% respond, what extrapolations can be made for the
  non-responding segment?
• To what extent will actual behavior differ from the attitudes expressed in
  the questionnaire?

The nonresponding group is not necessarily the same as the responders,
and hence a direct extrapolation of the data for the latter is inappropriate.
Indeed, the fact that one group responded and the other did not may
indicate significant differences in attitudes. Personal interviews with a
sample of the nonresponders might produce the desired information.

The other problem associated with questionnaires, the difference between the written attitude and the behavioral response, is difficult to
resolve. An individual might indicate in a questionnaire a willingness to
pay $100 for an item, but not follow through when confronted with the
actual decision. Possibly previous surveys on similar subjects can be
compared with subsequent behavior patterns to indicate the correlation
between the response and the actual behavior.

Specify Relationships Between Variables

Once the information has been gathered, it is then possible to write
equations, tables, or draw graphs relating the variables to each other (see
item 2 in Table 1.3). One approach is to use a regression analysis,[22]
wherein one assumes a functional relationship between variables contain-

ing undetermined coefficients. The coefficients are evaluated by minimizing an error function obtained from the data points and the assumed function. Alternatively, one can write continuity or rate equations as described in Chapter 2. The rate equations also contain undetermined coefficients, which are evaluated from the data.

It is not always possible to obtain analytic descriptions of the relationships between input and output variables. This is particularly true when the outputs are value judgments (worth estimates). In this case, the desired form of the input–output relationship may be a table where one column consists of descriptions of states of the system and another column consists of worth estimates. If, for example, one is concerned about noise aspects of a transportation network that includes highways, STOL aircraft and TACV, a worth estimate table might appear as in Table 1.4.[10] The worth estimate was based on a scale from 0 to 1, where 0 is the least desirable noise level and 1 is most desirable. Worth estimates are subjective and should reflect the opinions of the interested parties. The worth estimates for noise are then combined with worth estimates for other factors (cost, travel time, convenience, etc.) to obtain an overall worth for each modal mix, as discussed in Section 1.5.

Table 1.4
Noise Worth Estimates for a Transportation Network

Transportation Element	Decibel (dB) Noise Level	Distance of People from the Source	Worth Estimate
Freeway	68–73	100–400 ft	.6
	>73 (Noise continuous during rush hour)	<100 ft	.2
Airport	90–105	2–4 mi	.6
	>105 (Intermittant noise; exists during takeoff and landing)	>2 mi	.2
TACV	90–105	120–600 ft	.6
	>105 (Intermittant noise)	<120 ft	.2

Checking Relationships

Whenever possible, the input–output relationships should be checked (item 3, Table 1.3). This may be accomplished by predicting input–output values for a system in which these values are known or may be measured. There are many variables and many systems for which this cannot be done. If one is describing the behavior patterns of two opponents in a nuclear war situation, there is no previous example and it is not feasible to perform an experiment. There is usually a better opportunity for measuring relationships for physical parameters (noise, pollutant concentration, temperature, passengers per day, elapsed time, etc.) than for behavioral parameters. In general, the more global the system, the more difficult to check out because of the larger number of variables involved and the difficulty in isolating variables. When long time constants are involved (e.g., if it requires 20 years to observe some significant effects of a new transportation network), this presents an additional obstacle to verification.

Assumptions and Uncertainties

There is no problem that can be analyzed, whether in physics, economics, or sociology without some assumptions (item 4, Table 1.3). The velocity of a brick dropped from a height is influenced by the location of every star and the location of every person on earth. It is generally assumed, however, that the velocity of the brick may be calculated without considering these factors. Assumptions are often used to reduce the number of variables that need be considered so that the problem can be solved with the available facilities.

It is not always obvious whether or not an assumption is valid. Suppose a variable is assumed to remain constant, whereas in actuality it changes in a manner to produce a 20% fluctuation in the answer. If the problem is to obtain a better measure of the velocity of light in vacuum, the approximation is invalid because the velocity of light is presently known to a much greater degree of accuracy. If, however, the problem is to determine the number of people that will use a transportation network 20 years hence, it is probably a very good assumption. If an assumption simplifies a problem and the error introduced (if this error can be determined) is not significantly greater than the error introduced by other factors, such as the accuracy of the data, then the assumption is appropriate. The use of assumptions is an art that develops with experience.

A characteristic of an expert in a field is an ability to indicate the appropriate assumptions. Conflicting results from different analysts may be a consequence of different assumptions, and it is necessary to have an understanding of what assumptions have been made if discrepancies are to be resolved.

Assumptions that occur in environmental and social systems analysis may involve:

- The level of aggregation. Within any subgroup of the system we assume homogeneous behavior. Further subdivision can increase accuracy, but also increases complexity.
- The manner and place where boundaries are drawn around the problem. We might assume that factors outside the boundary remain unaltered by changes within the boundary, or that outside factors have little influence on the system.
- Structural relationships between elements of the system. Assumptions will be made concerning the manner in which one part of the system responds to changes in another part.
- Behavioral responses. It may be necessary to assume patterns of response by the interested parties to changes in the system.

Assumptions are often "best guesses" regarding properties of the system when we are uncertain as to the actual conditions that pervade. Uncertainties may exist concerning the validity of the data, projections of variables into the future, and structural and behavioral relationships. Uncertainty can be handled by a best guess, but this approach doesn't indicate the system response if the assumption is inappropriate (e.g., the best guess for a future population may be wrong). Other methods for dealing with uncertainty are:

- Uncertainty concerning the future may be reduced by waiting.
- Additional data may be collected so as to reduce data uncertainties. Prior to additional data collection it may be desirable to do a *sensitivity analysis*, in which the output variables are determined for different values of the uncertain inputs. If the outputs are relatively insensitive to variations in a given input, then it would not be necessary to obtain a more accurate determination for this variable.
- The systems analysis can be performed with a pessimistic estimate and an optimistic estimate, as well as with the best guess. This will indicate behavior under the possible range of values. An alternative may be selected that is relatively insensitive to change, which is not necessarily the most desirable alternative at the best guess value.

There is usually a cost associated with waiting, data collection, or running the model for additional values of the variables. These costs must be balanced against the benefits of reducing uncertainty.

1.5
EVALUATION

In the evaluation stage we attempt to provide a preferential ordering for the alternative solutions. At the present time, the most common method for alternative evaluation is based on the intuitive, visceral, and intellectual judgments of the decisionmaker. More often than not those variables that are easiest to quantify, as dollar cost, are given the greatest emphasis. If the decisionmakers are elected, they are often convinced that their intuitive responses represent the wishes of their constituency. This approach to problem solving has the advantages of being able to provide inexpensive, rapid responses to almost all problems. It has the disadvantages of not giving explicit consideration to the relative emphasis placed on the different facets of the system, it may not reflect the spectrum of interests in the community, and it may be susceptible to lobbyist pressures. One purpose of systems planning is to increase the proportion of intellectual to intuitive and visceral components of the decision process.

We shall be making a measurement different from many previous measurements we have made. It is straightforward to conclude that one object is longer than another by referring both lengths to a common scale of meters or fractions thereof. But how does one decide that it is more desirable to use one's resources to purchase two apples and three oranges or one apple, four oranges, and newspaper? The answer depends on the preferences of the individual and an optimum decision for one person may not be the same for another. In environmental and social systems planning, it is necessary to consider outcomes of our actions that cannot be measured in the same unit system, and this means that individual or group preferences must be incorporated into the evaluation procedure.

We constantly make choices between apples and oranges, usually without too much introspection, for our condition of well being is not drastically affected by exchanging an apple for an orange. However, if we consider changing jobs, investing our savings in a mutual fund, getting married, or selecting a course of study in the university, we will devote more time to the problem and make comparisons, consciously or not, between variables that, if they could be quantified, would be measured in different units. Environmental and social systems planning may involve millions of people, time periods measured in years, and large expenditures of funds. Our evaluation procedures for problems of this magnitude should be refined beyond the often hazy considerations used for individual action.

In the formulation stage a set of indices or attributes was developed to provide a means for determining the extent to which the goals are achieved. The systems approach to evaluation is to obtain a measure for the indices, and then to combine all of these individual measures so that the most desirable course of action may be selected. These two steps may be performed in a manner that is primarily qualitative or primarily quantitative, and both procedures will be discussed. Table 1.5 lists the steps in the evaluation stage.

Table 1.5
Evaluation

1.	Obtain a measure for each index or attribute.
2.	If more than one attribute is involved, combine the measures of step 1 into a single description or function from which an alternative may be selected.

Measurement of Attributes

The measure for each index (item 1, Table 1.5) may consist of a qualitative description. For example, in the transportation network system discussed in Section 1.3, a societal goal is to minimize detrimental effects on human physiology, such as pollution and accidents. A qualitative description for these attributes is given below for the two alternatives: (a) evolutionary expansion of the existing transportation system, and (b) the dual-mode system consisting of the addition of electric cars and buses. Statements similar to those in Table 1.6 would be given for each attribute.[10]

Some attributes have well-defined measurement scales, and one can progress toward quantification by calculating these values from the system model. In Fig. 1.8 a number of attribute scales are given, along with the value determined for each alternative.[10] (The scales for dB noise and baggage handling are linear, and the other scales are logarithmic.) For each attribute the more desirable outcome is on the right and the less desirable is on the left (e.g., higher speed is preferable, as is lower cost).

Table 1.6
A Qualitative Summary of Performance Attributes

Attribute	Evolutionary Expansion of Freeways	Dual-Mode System
Pollution	Direct air pollution is high in the vicinity of the right-of-way; moderate to high in the cities; moderate in the entire region. Large numbers of people affected by air and noise pollution in the cities. Indirect pollution is high in metropolitan areas because of the concentration of industry and population. Low pollution in the nonurban areas of the western portion of the region	Direct air, water, and noise pollution are low in all areas, with no cumulative effects. Indirect pollution moderate in dispersed cities, and low in greenbelt areas
Accidents	Accidents high in the vicinity of the right-of-way and on supporting street systems	Accidents moderate to low on guideway and at terminals. Accidents on other transport systems lowered because of the decrease in overall congestion

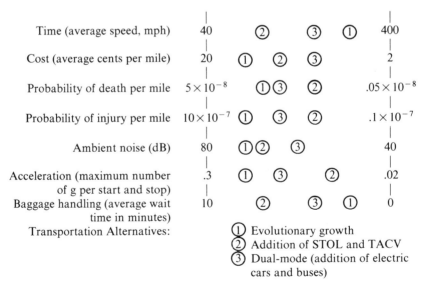

Time (average speed, mph)	40 ② ③ ①	400
Cost (average cents per mile)	20 ① ② ③	2
Probability of death per mile	5×10^{-8} ①③ ②	$.05 \times 10^{-8}$
Probability of injury per mile	10×10^{-7} ① ③ ②	$.1 \times 10^{-7}$
Ambient noise (dB)	80 ①② ③	40
Acceleration (maximum number of g per start and stop)	.3 ① ③ ②	.02
Baggage handling (average wait time in minutes)	10 ② ③ ①	0

Transportation Alternatives:
① Evolutionary growth
② Addition of STOL and TACV
③ Dual-mode (addition of electric cars and buses)

FIGURE 1.8. Attribute value for the different alternatives.

Utility Rating

The qualitative description and the calculation of attribute values for the measurement of goal indicators described in the previous paragraphs are objective results obtained from the analysis. The decision process involves the incorporation of subjective information as well (e.g., the desirability of having a transportation system that operates at 150 mph if it costs twice as much as a system that operates at 100 mph), and this may be accomplished by assigning to the attributes subjective worth estimates or utility ratings. The utility rating indicates the relative worth, to the individual, of a given performance characteristic.

In assigning a utility scale to an attribute it is useful to identify the extreme possibilities. The numbers given to the extreme outcomes do not affect the evaluation process and it is convenient, though not necessary, to give a value of unity to the best feasible result and a value of zero to the least acceptable outcome.

The best feasible result for each attribute might be obtained by defining the purpose of the project to be the fulfillment of that attribute. A least acceptable outcome would be the minimum value of the attribute that would make it acceptable to all interested parties. Reasonable extreme values for passenger atrributes on a specified link of a transportation network are given in Fig. 1.9.[10]

In most instances it is not possible to satisfy simultaneously the ideal outcomes for every attribute, since there is often a conflict between goals. A maximum performance criterion may not be consistent with a minimum time lapse or a minimum expenditure of funds. Also, an ideal may not result because of the uncertainty of events, such as a sudden population growth or a change in an environmental factor. A system may deteriorate or become obsolete and thereby fall below the ideal.

Since the system design does not generally result in a utility value of unity, it is necessary to assign a utility value to the partial fulfillment of a goal. The utility associated with partial satisfaction is determined from the subjective opinions of the decision maker and interested parties, and possible procedures for assigning utility u are:

1. A direct estimation of utility by the decision maker or interested party. For example, if a travel time of 30 min is considered ideal and 150 min is unacceptable, it might be agreed upon that 40 min would correspond to $u = .9$ and 50 min to $u = .8$. In this manner a curve of utility versus travel time is obtained. Further discussion of this approach is contained in the literature.[24]

	Ideal Outcome	Least Acceptable
Utility Rating	1	0

Attribute		
Travel time in minutes from origin terminal to destination terminal	30	150
Time in the terminal (min)	2	30
Trip cost (1970 $)	10	40
Probability of death $\times 10^7$	.01	10
Probability of injury $\times 10^7$	.1	100
Ambient noise (dB)	40	80
Emergency noise (dB) (1 sec. duration)	90	140
Acceleration in g	.02	.30
Jerk in g/sec.	.01	.25
Vibration in g	.01	.30
Temperature (°F)	67	64, 72
Pressure change (psi)	.01	.2
Seating (ft^3)	200	40
Privacy (number of other occupants in a 4′ radius)	0	8
Probability for being within ±2 min of schedule	.98	.80
Baggage-handling time (min)	0	10

FIGURE 1.9.   Extreme outcomes for passenger attributes.

2. Evaluation of utility ratings from indifference lotteries. This approach is discussed in Chapter 4.

Combining Index Measures

The selection of an alternative with specified values for the decision variables requires the reduction of the vector description for many attributes to a scalar. This may be done by a qualitative summary of the information presented in Table 1.6 and Fig. 1.8. Major differences and tradeoffs between alternatives would be indicated in such a summary. Some alternatives might be rejected on the basis of unacceptable performance on certain attributes. The final selection will usually require the decisionmaker to provide subjective assessments as to which performance properties he considers to be most significant.

Alternatively the utility ratings for each index may be used to generate a scalar that represents the overall utility for the entire system. This procedure deals explicitly with the value associated with the partial fulfillment of each attribute and also with the emphasis to be placed on one attribute relative to another.

The function that produces the scalar that is used to make the decision is termed the *objective function*. If, for example, cost is our only concern, then our objective function is the cost of the project, which we seek to minimize. That is, we choose the alternative that minimizes cost. (We have assumed that the utility function for cost is monotonically decreasing, so that minimization of cost is equivalent to maximization of utility.) If cost and time are important, the objective function would include both time and cost.

Weighting Attributes

There are a number of ways that attributes, or the utilities associated with attributes, may be combined into a single objective function. One possibility is to define the objective function F as

$$F = \mathbf{w} \cdot \mathbf{u} = \sum_{i=1}^{n} w_i u_i = w_1 u_1 + w_2 u_2 + \cdots + w_n u_n, \qquad (1.1)$$

where u_i is the utility associated with the ith attribute and w_i is a constant weighting factor assigned to u_i. Function F is the scalar product (dot product) of the vector $\mathbf{w}$ and the vector $\mathbf{u}$. In Eq. (1.1) it is assumed that the overall utility F can be represented as a linear, weighted summation of the utilities for the individual attributes.

A method for assigning weights is to rank-order the attributes in terms of perceived importance and then, considering pairwise comparisons of adjacent attributes, decide upon their relative importance. For example, if cost concerns are indicated to be first priority, time concerns are second, and safety concerns are third, one might decide that $w_1/w_2 = 1.5$, and

$w_2/w_3 = 1.2$, where w_1 is the weight for the utility of cost, w_2 is the weight for the utility of time, and w_3 is the weight for the utility of safety. Individual weights may then be determined by the addition of a normalization condition, such as $\Sigma_i w_i = 1$. For the previous example, this gives $w_1 = 0.45$, $w_2 = 0.30$, and $w_3 = 0.25$. Additional methods for deriving w_i are described in the literature.[23,24]

If the costs (or any other attributes) associated with the various alternatives are uncertain, so that cost is given by a probability distribution, there will be a corresponding probability distribution for the utility of cost. Therefore the objective function is also given by a probability distribution, and a reasonable strategy for selecting an alternative is to optimize the average value $\langle F \rangle$ for this function. However other strategies are available. For example, if one wished to avoid the worst cases, he could choose a strategy for which the probability of F falling below a specified number is minimized. (Further discussion of decision making under risk and uncertainty is presented in Chapter 4.)

An Objective Function With Constraints

In addition to combining attributes by the method given by Eq. (1.1), another procedure is to place fixed constraints on all variables but one, and then optimize the objective function for that one remaining variable. For example, if cost and travel time are the relevant attributes, one might impose a maximum travel time and then choose the alternative with the minimum cost that does not exceed the specific time. Alternatively, a maximum cost may be indicated and travel time would then be minimized. If the project has a fixed budget, the latter approach might be the more reasonable one to select.

Equivalence Relationships

An approach that seeks to be more precise in combining attributes, but requires more effort than the two previous approaches, is to define equivalence relationships between attributes or between the utilities associated with the attributes. For example, if it were possible to determine the increase in fare that a passenger would be willing to pay for a reduction in travel time, a "cost equivalent" of travel time could be established. This cost equivalence $c_{eq}(t)$ is added to the actual passenger cost to yield a total cost objective function that is to be minimized.

Similarly, an objective function can also be obtained by establishing equivalence relationships between the utility functions for the various attributes. Equation (1.1) is a special case of this method where the equivalence is independent of the utility values and so a straight-line relationship holds. From (1.1) one finds that if u_i is changed with respect to a change in u_j in the ratio of the weights $(-w_j/w_i)$, the objective function F is unaltered. This means that for a specified increase in u_j an individual remains equally satisfied if u_i is increased by the change in u_j times the factor (w_j/w_i). Figure 1.10 shows the lines of equal satisfaction for the linear case exemplified by Eq. (1.1). These lines are *indifference curves*, for the interested party is indifferent to being located at any point along a given line. Each curve in Fig. 1.10 represents a different level of satisfaction, and the direction of maximum increase in total utility is orthogonal to the lines of constant utility (as shown by the arrow). The direction of maximum change in utility is of interest because it indicates how one should move to achieve the most rapid improvement. This direction is given by the direction of ∇F, where ∇ is the gradient operator:

$$\nabla F = \sum_i \mathbf{l}_i \frac{\partial F}{\partial u_i},\qquad(1.2)$$

where $\mathbf{l}_i$ is the unit vector in the u_i direction. For $F = w_i u_i + w_j u_j$, Eq. (1.2)

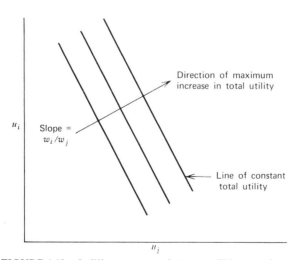

FIGURE 1.10. Indifference curves between utilities u_i and u_j.

gives

$$\nabla F = \mathbf{1}_i \frac{\partial F}{\partial u_i} + \mathbf{1}_j \frac{\partial F}{\partial u_j} = \mathbf{1}_i w_i + \mathbf{1}_j w_j. \tag{1.3}$$

Therefore the vector direction for maximum total utility increase has a projection w_i along the u_i direction and a projection w_j along the u_j direction, with a corresponding slope of (w_i/w_j) in the $(u_i - u_j)$ plane.

Cost and Effectiveness

Combination of attribute procedures may be illustrated by considerations of cost and effectiveness. This is a two-attribute problem, where effectiveness refers to any performance characteristic, or combination of performance characteristics, of the system. In a transportation network, effectiveness might be a worth estimate for the system excluding the cost factor, or it might refer to a single factor such as travel time or safety. For a given alternative, the cost and effectiveness values are calculated as the control (i.e., decision) variables are changed. From these calculations, a curve of effectiveness as a function of cost may be plotted (see Fig. 1.11). These curves are typical in that there is an initial increase in effectiveness with increased cost, but effectiveness tends to saturate at high cost. For instance, continued cost increases will not produce comparable decreases in travel time; other factors, such as technology, tend to limit the minimum time that can be achieved.

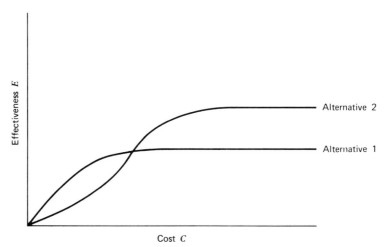

FIGURE 1.11.   Effectiveness as a function of cost for two alternatives.

The two attributes, cost and effectiveness, are measured in different units. Cost is measured in dollars. If effectiveness is a worth estimate it is nondimensional, and if it is a travel time it is measured in minutes, and so on. A selection of one alternative requires that the two-dimensional vector description of each alternative, cost and effectiveness, be reduced to a scalar. Let us consider several previously suggested procedures: to establish a constraint on one variable and optimize the other, or utilize an indifference relationship between variables.

If a minimum level of effectiveness E_{min} is given, and cost is to be minimized, the alternative that is selected depends on the value for E_{min}. As shown in Fig. 1.12, for E_{min_1} the minimum cost results from alternative 1, whereas for E_{min_2} only alternative 2 is acceptable. If maximum cost C_{max} is specified, and effectiveness is to be maximized, the alternative selection again depends on the value for C_{max}. For C_{max_1}, alternative 1 is preferable, and for C_{max_2}, alternative 2.

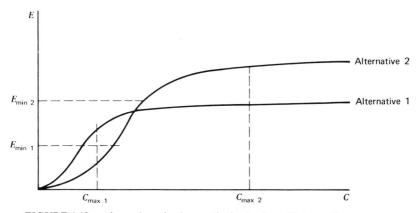

FIGURE 1.12.   Alternative selection on the basis of specification of a constraint.

It is often suggested that alternative selection should be based on maximization of the effectiveness-to-cost ratio, for this yields the largest effectiveness per dollar. Several lines of constant E/C are superimposed on the cost–effectiveness curves in Fig. 1.13, with the arrow indicating the direction of increasing E/C. From this figure we see that the maximum E/C is attained at point A, where a line of constant E/C is tangent to alternative 1. If the approach is to maximize E/C, this means that one is equally satisfied for a specified E/C regardless of the values for E and C.

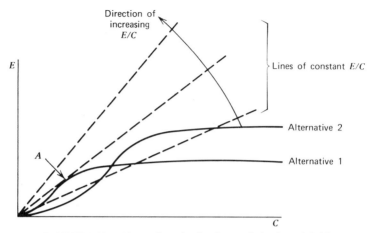

FIGURE 1.13.   Alternative selection by maximization of E/C.

Therefore lines of constant E/C are lines of indifference or constant total utility. Therefore maximization of E/C is equivalent to maximization on the basis of establishing an indifference relationship which says that satisfaction is constant for a cost increment that results in a proportional increment in effectiveness.

A more reasonable type of indifference curve might be similar to the dashed lines in Fig. 1.14, where equal satisfaction requires larger increments in effectiveness for a given cost increment as the cost increases. For

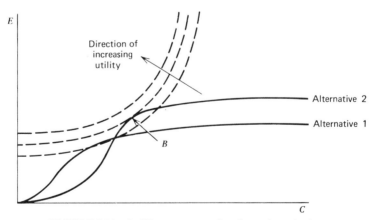

FIGURE 1.14.   Indifference curves for alternative selection.

the lines as drawn in Fig. 1.14, point B provides the maximum utility. We see from this cost–effectiveness discussion that there are a number of different strategies for making a decision, and in Chapter 4 we shall explore the decision process in greater detail.

The procedures discussed in this chapter are intended primarily as a point of initiation, rather than as a rigid framework for the investigation of a problem. Each situation is unique, so that the format must be modified as the analysis progresses. The reader is referred to Appendix A for a description of some of the practical problems encountered in a systems analysis of health-care delivery in San Mateo County, California.

REFERENCES

1. C. W. Churchman, K. L. Ackoff, and E. L. Arnoff, *Introduction to Operations Research*, Wiley, New York (1957).

2. H. Chestnut, *Systems Engineering Tools*, Wiley, New York, (1966), p. 8.

3. H. A. Affel, Jr., "System engineering," *Int. Sci. Technol.* **18**, (Nov. 1964).

4. R. A. Howard, "The Foundations of Decision Analysis," IEEE Trans. on Systems Science and Cybernetics, *SSC*-4, 211 (Sept. 1968).

5. D. W. North, "A Tutorial Introduction to Decision Theory," IEEE Trans. on Systems Science and Cybernetics, *SSC*-4, 211, (Sept. 1968).

6. H. Raiffa, *Decision Analysis, Introductory Lectures on Choices Under Uncertainty*, Addison–Wesley, Reading, Mass. (1968), pp. 295–297.

7. E. S. Quade and W. I. Boucher (eds.), *Systems Analysis and Policy Planning, Applications in Defense*, Amer. Elsevier, New York (1968), pp. 15–17.

8. J. H. Seinfeld and C. P. Kyan, "Determination of Optimal Air Pollution Control Strategies," *Socio-Econ. Plann. Sci.* **5**, 173–190 (1971).

9. D. J. Rose, J. H. Gibbons, and W. Fulkerson, "Physics Looks at Waste Management," *Phys. Today* **25** (2), 32–41 (Feb. 1972).

10. F. S. Pardee, C. T. Phillips, and K. V. Smith, *Measurement and Evaluation of Alternative Regional Transportation Mixes*, Vols. 1–3, Rand Corp., RM-6324-DOT (Aug. 1970).

11. H. B. Wolfe and M. L. Ernst, "Simulation Models and Urban Planning," Ch. 3 in P. M. Morse and L. W. Bacon (eds.), *Operations Research for Public Systems*, M.I.T. Press, Cambridge, Mass., 3rd print. (1971).

12. M. H. Schussman, *Goalsmanship (Developing Educational Philsophy, Goals, and Objectives)*, report prepared for San Mateo County School Boards Assoc., Redwood City, Calif. (Sept. 1, 1970).

13. F. D. Kennedy, "Development of a Community Health Service System Simulation Model," IEEE Trans. on Systems, Science and Cybernetics, *SSC*-5, 204 (July 1969).

14. W. J. Horvath, "Operations Research in Medical and Hospital Practice," Ch. 6 in P. M. Morse and L. W. Bacon (eds.), *Operations Research for Public Systems*, M.I.T. Press, Cambridge, Mass., 3rd print. (1971).

15. A. Blumenstein and R. C. Larson, "A Systems Approach to the Study of Crime and Criminal Justice, Ch. 7 in P. M. Morse and L. W. Bacon (eds.), *Operations Research for Public Systems*, M.I.T. Press, Cambridge, Mass., 3rd print. (1971).

16. I. R. Hoos, *Systems Analysis in Public Policy. A Critique*, Calif. U.P., Berkeley (1972).

17. K. J. Arrow, *Social Choice and Individual Values*, Wiley, New York (1951).

18. A. Etzione and E. W. Lehman, "Some Dangers in 'Valid' Social Measurement," *Ann. Amer. Acad. Polit. Soc. Sci.*, **1**, 1–15 (May 1967).

19. O. Helmer and N. Rescher, "On the Epistemology of the Inexact Sciences," *Manage. Sci.* **6**, 1 (Oct. 1959).

20. B. Brown and O. Helmar, "Improving the Reliability of Estimates Obtained from a Consensus of Experts," Rand, p2986 (Sept. 1964).

21. M. G. Weiner, "Gaming," Ch. 7 in E. S. Quade and W. I. Boucher (eds.), *Systems Analysis and Policy Planning, Applications in Defense*, Amer. Elsevier, New York (1968).

22. A. M. Mood, *Introduction to the Theory of Statistics*, McGraw-Hill, New York (1950), Ch. 13.

23. R. J. Miller, III, *Assessing Alternative Transportation Systems*, Rand, RM-5865-DOT (Apr. 1969).

24. R. L. Ackoff, *Scientific Method Optimizing Applied Research Decisions*, Wiley, New York (1962), pp. 76–93.

PROBLEMS

1.1 A study by the Urban Institute of Washington, D. C. attempted to measure the quality of life in 18 urban areas in the United States. Before reading further, define the attributes you would use to measure quality of life in the city. For each attribute indicate the method of measurement.

The actual attributes are listed as follows, with the method of measurement in parentheses: unemployment (% unemployed), educational attainment (median school years completed), poverty (% low-income households), health (infant-mortality rate), mental health (reported suicide rate), racial equality (unemployment ratio between nonwhite and white), air quality (pollutant concentration), income level (adjusted per capita income), transportation (average annual transportation cost to a moderate-income family of four), housing (annual cost to family of four), public order (robberies per 10^5 population), community concern (United Fund donations per capita), social disintegration (drug addicts per 10^4 population), and citizen participation (presidential voting rate). Discuss differences between your list and the above list. Did you include open space, cultural facilities, and climate on your list? How would each of these attributes be measured?

1.2 Eighteen urban areas were ranked on a scale from 1 to 18 (1 corresponds to the most desirable ranking) with regard to each of the 14 attributes listed in Problem 1.1. These rankings are shown in the chart on the following page.

Suppose you wish to select a residence based on the information in this chart.

(a) What city (or cities) would you choose if you selected the city with the best ranking for its poorest attribute? For example, New York ranks eighteenth in

	Unemployment	Poverty	Income	Housing	Health	Mental Health	Public Order	Racial Equality	Community Concern	Citizen Participation	Education	Transportation	Air Quality	Social Disintegration
San Francisco Bay Area	16	17	2	14	2	18	13	2	13	7	2	16	1	5
New York	9	9	4	17	9	1	18	1	18	14	9	1	10	7
Los Angeles–Long Beach	18	15	3	10	2	17	11	3	17	11	3	7	8	3
Chicago	2	5	5	13	18	3	14	6	15	2	7	13	16	1
Philadelphia	7	9	14	9	17	13	6	9	10	8	13	2	15	NA
Detroit	17	3	1	5	12	15	17	8	7	5	13	8	14	NA
Washington	1	2	8	11	5	7	15	7	16	18	1	10	3	6
Boston	4	1	17	18	6	10	3	NA	12	2	6	15	5	4
Pittsburgh	14	12	13	3	11	11	5	5	2	2	13	3	17	NA
St. Louis	10	13	12	8	10	4	9	11	6	11	17	14	18	2
Baltimore	5	11	16	4	15	14	16	4	14	14	18	9	11	NA
Cleveland	12	7	9	15	8	16	10	12	1	9	9	6	13	NA
Houston	5	17	11	1	16	12	12	10	11	16	9	17	5	NA
Minneapolis–St. Paul	14	4	6	6	1	5	7	NA	5	1	3	11	2	NA
Dallas	3	16	7	2	14	6	8	NA	9	16	3	4	3	NA
Milwaukee	10	5	15	16	4	8	1	NA	8	5	7	4	11	NA
Cincinnati	7	14	10	7	7	9	2	NA	3	11	13	12	9	NA
Buffalo	12	8	18	12	13	2	4	NA	4	10	9	18	5	NA

NA—not available

public order, which is its poorest attribute, whereas Philadelphia ranks seventeenth in health, its poorest attribute. Therefore on this basis Philadelphia would be chosen over New York. [Note: When data are not available, (NA), all rankings are equally probable, so that a reasonable ranking is at the median; see Section 4.3 for further discussion.]

(b) Which city (or cities) would you choose if you selected on the basis of the highest number of top rankings?

(c) Select the five attributes that you consider to be most important. Rank order these and weight them by the method described in Section 1.5 under the heading *Weighting attributes*. Assign utility values to the rankings with first ranking corresponding to unity utility and eighteenth ranking corresponding to zero. Use a linear utility relationship for intermediate rankings. Select an urban area based on a weighted utility sum, as given by Eq. (1.1).

Discuss the advantages and disadvantes of each method for choosing a city.

1.3 An elementary school district is confronted with a declining enrollment and a projected budget deficit. A course of action is to be selected. List the concerns of each of the following interested parties:

- Pupils
- Parents of pupils
- Residents in the school district without children in the schools
- Teachers
- School administrators (superintendant, principals)
- School board (a five-member elected body responsible for policy decisions)

Specify a measure for gauging the status of each of the concerns. [See R. F. Mager, *Preparing Instructional Objectives*, Fearon, Belmont, Calif. (1962).]

Alternative courses of action for responding to the deficit are:

- Have a bond issue.
- Eliminate extracurricula activities (e.g., music program, counseling, and excursions).
- Dismiss teachers and enlarge class size.
- Close one school in the district and bus students to the remaining schools, eliminating maintenance of one facility.
- Combine with another school district to reduce administrative overhead.

For each alternative, rate the concerns on a scale of one to three (using your best estimate, and giving reasons), with one corresponding to the most desirable outcome. Which alternative would you choose as a member of the school board? Why?

1.4 Data have been collected relating the weight of a child to his performance on a spelling test. If a linear functional relationship is assumed

$$S = aW + b$$

where S = score and W = weight, determine a and b by regression. That is, minimize the mean value of the square of the error between the assumed function and the data points. If S_i, W_i is the score and weight, respectively, for the ith child, show that this procedure gives

$$a = \frac{n \sum_{i=1}^{n} S_i W_i - \left(\sum_{i} S_i \right) \left(\sum_{i=1}^{n} W_i \right)}{D}$$

$$b = \frac{\left(\sum_{i=1}^{n} W_i^2 \right) \left(\sum_{i=1}^{n} S_i \right) - \left(\sum_{i=1}^{n} S_i W_i \right) \left(\sum_{i=1}^{n} W_i \right)}{D}$$

where n = number of children, and

$$D = n \sum_{i=1}^{n} W_i^2 - \left(\sum_{i=1}^{n} W_i \right)^2$$

1.5 Rate each of the following energy sources: (a) solar, (b) fusion reactors, (c) fission reactors, (d) geothermal, (e) hydroelectric, (f) natural gas, (g) coal, (h) petroleum, and (i) wind power with regard to the following attributes: (a) safety, (b) resource allocation necessary for bringing to market, (c) pollution, and (d) abundance. Use a scale of 1–3 (1 being most desirable), and give a brief (one or two sentences) reason for your ranking [see D. J. Rose, Energy policy in the U. S., *Sci. Amer.*, **230** (1), 20–29, (Jan. 1974); E. S. Cheny, U. S. Energy Resources, *Amer. Sci.* **62**, 1, 14–22, (Jan.–Feb. 1974)].

1.6 If indifference lines in the effectiveness (E)–cost (C) plane are given by

$$\frac{E}{C} = \text{constant},$$

by use of the gradient operator determine the curves that correspond to the paths for maximum increase in utility. What type of figure are these curves?

1.7 If indifference lines in the E–C plane are given by

$$E(C_0 - C)^2 = \text{constant for } C < C_0,$$

where C_0 is a positive constant: (a) draw the indifference lines and (b) from the gradient operator, determine an equation relating E and C that specifies the paths for maximum increase in utility.

1.8 Suppose you wish to know whether or not there will be a shortage of engineers and physicists within the next decade. Outline the method you might use for making this projection, and specify the important assumptions and uncertainties [see A. M. Cartter, "Scientific manpower for 1970–1985," *Science* **172**, 132–140, (*Apr.* 9, 1971); Science and Engineering Doctorate Supply and Utilization 1968–1980, *NSF* 69-37 (1969); L. Grodzins, "*The Manpower Crisis in Physics*," Bull. Amer. Phys. Soc. **16**, 737–749 (*June* 1971); W. R. Brode, *Manpower in science and engineering, based on a saturation model*, Science **173**, 206–213 (*July* 16, 1971)].

Chapter Two □ Continuity Equations

2.1
INTRODUCTION

The goal of modeling, as discussed in Chapter 1, is to relate output variables to input variables, so that the consequences of the imposition of alternative policies may be predicted. An important part of the modeling process, once the problem has been structured, is to identify possible continuity and rate of change relationships. Namely, the flowrate of a variable into and out of any part of the system is related to the rate of change of the level of the variable; that is, everything has to go somewhere. Conservation of monetary flows is the function of accounting in a business organization, where the rate of monetary input is equated to the rate of cash accumulation plus the rate of expenditure.

In this chapter, some of the principles of continuity equations are presented. The discussion of each topic is introductory rather than exhaustive, for the material is designed to provide an overview of a wide variety of problems and to illustrate similarities between systems.

In ecosystems, a state variable of interest may be the number or mass of a given species. Figure 2.1 illustrates the various inputs and outputs that affect the mass of a species. The only source of mass increase is consumption. Birth, which gives an increase in number, does not represent an increase in mass of the species, for the combined mass of parent and offspring immediately after birth cannot exceed the mass of the parent (including the fetus) before birth. Indeed, there is usually waste material produced at birth that results in a mass loss to the species, and is designated as one of the output flows shown in Fig. 2.1.

Mass reduction rate due to death, either natural or from an encounter with an enemy, is generally a function of the species mass. If the species mass increases, the death mass loss rate also tends to increase (e.g., if the

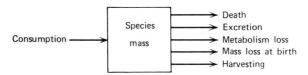

FIGURE 2.1.   Input and output flows affecting the total mass of a species.

deathrate per year is $1/1000$, then if the population is doubled the number of deaths per year also doubles). Mass loss from harvesting (e.g., species cultured to provide food for man) will usually have a different functional dependence than other causes of death and, therefore, is listed as a separate loss item. The remaining mass-loss factors result from excretion and from metabolic processes that convert mass to energy.

The conservation of mass equation (or mass-balance equation) for Fig. 2.1 is:

Mass increase per unit time = (mass consumption per unit time)

$$- \text{(mass loss per unit time due to death)}$$

$$- \text{(mass loss per unit time due to excretion)}$$

$$- \text{(mass loss per unit time due to metabolism)}$$

$$- \text{(mass loss per unit time due to births)}$$

$$- \text{(mass loss per unit time due to harvesting)}.$$

$$(2.1)$$

The next step is to determine the functional form for each of the terms on the right-hand side of (2.1). Frequently, one takes the first few terms in a power series expansion to represent each flow term, as will be illustrated by the examples presented in the remaining sections of this chapter.

2.2
RATE PROCESSES FOR A SINGLE GROUP OR SPECIES

Differential Equations

Let us first consider the problem where all members of the group or all particles of interest are identical. If we are counting the number of members N at any time t, the rate of change of N with respect to time

(dN/dt) will be given by

$$\frac{dN}{dt} = (\text{growth in numbers per unit time})$$

$$- (\text{loss in numbers per unit time}). \quad (2.2)$$

In general, the growth per unit time $G(N,t)$ is a function of the number N and time, and the loss per unit time $L(N,t)$, is similarly a function of N and t. If growth results from births (rather than migration, mutation, or creation), and loss results from deaths, G and L are often written as

$$G(N,t) = (\text{probability of a birth per unit time}$$

$$\text{per individual}) \times N. \quad (2.3)$$

$$L(N,t) = (\text{probability of a death per unit time}$$

$$\text{per individual}) \times N. \quad (2.4)$$

The probability of a birth or death per unit time per individual may be a function of time and is often a function of N. For example, as the number increases, the food consumption may decrease and lead to a lower birth-rate. Alternatively, as N increases there may be more male–female encounters per unit time leading to a higher birthrate.

Initially, we shall consider birth- and deathrates to be independent of N and t, and subsequently we shall generalize to population-dependent rates. In the case of birth and deathrates independent of N and t, the applicable differential equation is

$$\frac{dN}{dt} = aN - bN = \lambda N, \quad (2.5)$$

where a is the probability of a birth per unit time per individual, b is the probability of a death per unit time per individual, $\lambda = a - b$, and a, b, and λ are constants. The solution to (2.5) is

$$N = N(0)e^{\lambda t}, \quad (2.6)$$

where $N(0)$ is the population at $t = 0$. If a, b, and therefore λ are functions of time, the solution is

$$N = N(0) \exp \int_0^t \lambda(t)\, dt. \quad (2.7)$$

If $\lambda < 0$ (i.e., the deathrate exceeds the birthrate), the population eventually decreases to zero, if $\lambda > 0$ the population becomes infinite, and for $\lambda = 0$ the population remains constant.

If λ is initially positive, growth will eventually be limited by N-dependent birth- and deathrates. We shall approximate the N dependence of a and b by considering the first two terms in the power series expansion

$$
\left.
\begin{aligned}
a &= a_0 - a_1 N \\
b &= b_0 + b_1 N
\end{aligned}
\right\}
\tag{2.8}
$$

where, for a_0, b_0, a_1, $b_1 \geqslant 0$, we have assumed that the birthrate decreases with N and the deathrate increases with N. Substitution of (2.8) into (2.5) yields

$$
\frac{dN}{dt} = (a_0 - b_0)N - (a_1 + b_1)N^2
\tag{2.9a}
$$

$$
= c_0 N - c_1 N^2,
\tag{2.9b}
$$

where c_0 and c_1 are constants.

We arrived at the term $c_1 N^2$ by considering N-dependent birth- and deathrates, but the same term would result from lethal encounters between members of the group meeting two at a time. That is, the probability of a death per unit time per individual is proportional to N, since increasing N increases the likelihood of an encounter in any given time interval. Therefore the rate of change of population per unit time resulting from an encounter is proportional to N^2.

The solution to (2.9) is

$$
N = \frac{c_0/c_1}{1 + \left[\dfrac{c_0/c_1}{N(0)} - 1 \right] \exp(-c_0 t)}.
\tag{2.10}
$$

If $c_0 < 0$, N decays to zero with time as shown in Fig. 2.2a. If $0 < c_0 < c_1 N(0)$, N decays to the value (c_0/c_1) as shown in Fig. 2.2b. If $c_1 N(0) < c_0$, N grows to the value (c_0/c_1) as illustrated in Fig. 2.2c. Several species have been observed to exhibit the growth curve in Fig. 2.2c.

Equation (2.9) is an approximation for any actual growth condition, and the validity of (2.9) depends on the relative magnitude of the neglected terms in comparison to the terms that have been retained. For small

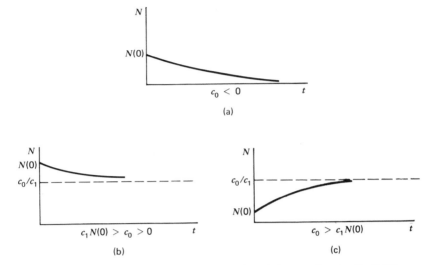

FIGURE 2.2.   The behavior of N as a function of time, as given by Eq. (2.10).

changes in N about an equilibrium value, it is a good approximation to use only the first few terms in the power series expansion about the equilibrium condition. For large changes in N it may be necessary to include higher-order terms. A further discussion of population growth and decay is given in Chapter 11 of Watt's book.[1]

Difference Equations

Equation (2.9) is a first-order differential equation, in which it is assumed that N is a continuous function of time. It is not uncommon to express changes in a variable as a difference equation in time, where time is divided into discrete intervals. There are a number of reasons for using difference equations rather than differential expressions:

1. The available data may have been gathered at discrete intervals, so that changes are measured in steps.
2. The variable may actually change in a discrete manner. Dollar flow into and out of an organization may occur on a weekly basis. Energy changes in atoms, molecules and force fields are hypothesized to occur in discrete jumps.
3. A solution of the equation by the use of a digital computer necessitates writing a difference equation.

If we choose our time interval to be Δ, then the difference equation for a population that changes because of births and deaths is

$$N(t+\Delta) - N(t) = a\Delta N(t) - b\Delta N(t), \tag{2.11}$$

where, as before, a is the probability of a birth per unit time per individual. Therefore $a\Delta$ is the probability of a birth per individual in time Δ, and $a\Delta N$ is the number of births in time Δ. As $\Delta \to 0$, (2.11) reverts to (2.5).

Stochastic Processes

For both the differential form of (2.9) and the difference form of (2.11), function N is *deterministic*. That is, an exact value for N may be calculated given the initial conditions. This result follows from the assumption that births and deaths occur at a specified rate with complete certainty. But births and deaths are discrete, random events, and one is constrained to specifying that at a time t there is a probability $P_t(N)$ that the population is N. It is possible (though not likely), for example, to have 10 deaths without an intervening birth, even though the birth and death probabilities may be comparable. When variables are expressed in terms of a probability function, rather than having a definite value, we say that the variables are *stochastic*. In this section we wish to determine the population as a function of time when births and deaths are considered to be random events described by a probability. In particular, we want to observe any differences between the deterministic and stochastic approaches.

Let us first consider some general properties of stochastic processes and then relate these properties to the birth–death problem specified by (2.8) and (2.9).

If the population number is J at time t, we define the system as being in state J at time t. Let

$$P(I,J) = \text{probability that the system is in state } J$$
$$\text{at time } t+\Delta, \text{ given that the system is}$$
$$\text{in state } I \text{ at time } t.$$

If the $P(I,J)$ depend solely on the properties of states I and J, and do not depend on the path or manner in which state I was reached, the process is termed a *Markov chain*.[2] A Markov chain is said not to have a memory, for the probability of going from I to J does not depend on the condition of the system prior to reaching state I. We shall see that the birth–death problem considered previously is a Markov chain.

The probability of being in state N at time $t+\Delta$, $P_{t+\Delta}(N)$, is the probability of being in state I at time t multiplied by the probability of going from state I to state N, and summed over all possible initial states:

$$P_{t+\Delta}(N) = P_t(0)P(0,N) + P_t(1)P(1,N) + \cdots + P_t(N)P(N,N) + \cdots$$

$$= \sum_I P_t(I)P(I,N). \tag{2.12}$$

It is convenient to express (2.12) in matrix notation, where P_t is a row vector

$$P_t = (P_t(0) \quad P_t(1) \quad P_t(2) \cdots),$$

and P is a matrix

$$P = \begin{bmatrix} P(0,0) & P(0,1) & P(0,2) & \cdots \\ P(1,0) & P(1,1) & P(1,2) & \cdots \\ P(2,0) & P(2,1) & P(2,2) & \cdots \\ \vdots & \vdots & \vdots & \end{bmatrix}.$$

In matrix notation, (2.12) becomes

$$P_{t+\Delta} = P_t P, \tag{2.13}$$

where $P_{t+\Delta}$ is the row vector

$$P_{t+\Delta} = [P_{t+\Delta}(0) \quad P_{t+\Delta}(1) \quad P_{t+\Delta}(2) \cdots].$$

If we substitute $t+\Delta = m\Delta$ into (2.13), we find that

$$P_{m\Delta} = P_{(m-1)\Delta}P = P_{(m-2)\Delta}P^2 = P_0 P^m. \tag{2.14}$$

Equation (2.14) expresses the probability vector after m steps in time as a function of the initial occupation probabilities P_0.

The sum of the elements of any row of the P matrix equals unity. That is,

$$\sum_N P(I,N) = 1. \tag{2.15}$$

Equation (2.15) is a statement of the fact that the probability of making a transition from the Ith state to any other state, including remaining in the Ith state, is unity, A matrix for which the sum of the row elements is unity is termed *stochastic*. The P^m matrices are also stochastic since the probability of transfer from one state to any other in time $m\Delta$ is unity.

Let us now consider the birth–death problem given by Eq. (2.9). If we choose the time interval Δ to be sufficiently small so that there is a negligible probability of having more than one event in the interval, the only possibilities during Δ are a single birth, a single death, or neither a birth nor a death. Therefore, we need to consider only three element types in the P matrix: $P(N-1,N)$, $P(N,N)$, and $P(N+1,N)$. These elements are

$$P(N-1,N)=\text{probability of a birth in the time}$$
$$\text{period } \Delta, \text{ given that there are}$$
$$N-1 \text{ at time } t$$

$$= [a_0 - a_1(N-1)]\Delta(N-1) \tag{2.16}$$

$$P(N,N)=\text{probability that there is neither}$$
$$\text{a birth nor a death in time period}$$
$$\Delta, \text{ given that there are } N \text{ at time } t$$
$$= 1-(\text{probability of a birth or a death})$$

$$= 1-[(a_0-a_1N)\Delta N+(b_0+b_1N)\Delta N] \tag{2.17}$$

$$P(N+1,N)=\text{probability of a death in time period } \Delta$$
$$\text{given that there are } N+1 \text{ at time } t$$

$$= [b_0 + b_1(N+1)]\Delta(N+1) \tag{2.18}$$

We note that in calculating (2.16) through (2.18) we did not need to include any information about the system prior to time t, and therefore this is a Markov chain process. Also, the $P(I,J)$ are not explicit functions of time, and so the chain is said to be *homogeneous in time*.

To consider a specific example, let

$$a_0 = 1 \text{ s}^{-1} \qquad\qquad b_0 = .4 \text{ s}^{-1}$$
$$a_1 = .9\times10^{-2} \text{ s}^{-1} \qquad b_1 = .3\times10^{-2} \text{ s}^{-1}$$

and $\Delta = .05$ s. From (2.16) through (2.18) we have that the P matrix is

$$P = \begin{pmatrix} 1 & 0 & 0 & 0 & 0 & 0 & 0 & \cdots \\ .02 & .93 & .05 & 0 & 0 & 0 & 0 & \cdots \\ 0 & .04 & .86 & .10 & 0 & 0 & 0 & \cdots \\ 0 & 0 & .06 & .79 & .15 & 0 & 0 & \cdots \\ 0 & 0 & 0 & .08 & .73 & .19 & 0 & \cdots \\ 0 & 0 & 0 & 0 & .10 & .66 & .24 & \cdots \\ \vdots & \vdots & \vdots & \vdots & \vdots & \vdots & \vdots & \end{pmatrix} \qquad (2.19)$$

If we start the system out with $N = 2$ so that

$$P_0 = (0 \quad 0 \quad 1 \quad 0 \quad 0 \quad 0 \quad \cdots)$$

then at $t = .05$ s

$$P_{.05} = P_0 P = (0 \quad .04 \quad .86 \quad .10 \quad 0 \quad 0 \quad \cdots)$$

and at $t = .1$ s

$$P_{.1} = P_{.05} P = (.0008 \quad .072 \quad .75 \quad .16 \quad .015 \quad 0 \quad 0 \quad \cdots)$$

We note that at 0.1 s there is a nonzero probability that $N = 0$. This result is in conflict with the deterministic equation (2.10), which exhibits the growth curve shown in Fig. 2.2c. The stochastic analysis indicates that the population may be extinguished, whereas the deterministic does not.

Since we have neglected multiple events during Δ, the smaller we make our time interval Δ the more accurate are the calculations. However the smaller is Δ, the more steps are required to span any given time interval. A reasonable upper limit for Δ is when the contribution to the transition probabilities from terms that have Δ as a coefficient are small compared to the terms without Δ. In the birth–death example, this means that the diagonal terms in P should be large compared to the off-diagonal terms. From (2.16) through (2.18) we see that for large N values, Δ should be reduced to maintain comparable accuracy.

The deterministic equation (2.9) reaches a steady-state value N_s obtained by setting $(dN/dt) = 0$:

$$N_s = \frac{a_0 - b_0}{a_1 + b_1}. \qquad (2.20)$$

Using the values selected for a_0, a_1, b_0, and b_1, we find that $N_s = 50$. From

(2.16) and (2.18) we have for $N = 50$,

$$P(50,51) = P(50,49) = 27.5\Delta. \tag{2.21}$$

Thus when $N = 50$, the probability of a birth equals the probability of a death and the population tends to stay around this value. For $N < 50$, the probability of a birth exceeds the probability of a death and so the population is most likely to grow. For $N > 50$, the probability of a death exceeds the probability of a birth and so the population is most likely to diminish. Thus, both the stochastic and deterministic equations give the same steady-state mean value.

From a table of random numbers[3] (this table is reproduced for Problem 2.1) we may simulate the random birth–death process. If, for example, we start with $N = 2$, then from (2.19) we have that $P(2,1) = 0.04$, $P(2,2) = 0.86$, and $P(2,3) = 0.10$. If the random number is from 0.01 to 0.04, we then make $N = 1$ at $t = 0.05$ seconds; if the random number is from 0.05 to 0.90, then $N = 2$ at $t = 0.05$ s, and if the number is from 0.91 to 0.00 then $N = 3$ at $t = 0.05$ s. In this manner, the probability of arriving at a new state is the same as that specified by the $P(I,J)$. Following the random numbers across a row, as listed in Problem 2.1, the population as a function of time is given in Table 2.1. Figure 2.3 is a plot of N as a function of time for both the deterministic and stochastic equations.

Table 2.1
Population as a Function of Time, Using a Table of Random Numbers

N	t in s	$P(N, N-1)$	$P(N, N)$	$P(N, N+1)$	Random Number
2	0.00	.04	.86	.10	03
1	.05	.02	.93	.05	47
	.10				43
	.15				73
	.20				86
	.25				36
	.30				96
2	.35	.04	.86	.10	47
	.40				36
	.45				61
	.50				46
	.55				98
3	.60	.06	.79	.15	63
	.65				71
	.70				62
	.75				32

Table 2.1   (*Continued*)

	.80				26
	.85				16
	.90				80
	.95				45
	1.00				60
	1.05				11
	1.10				14
	1.15				10
	1.20				96
4	1.25	.08	.73	.19	97
5	1.30	.10	.66	.24	74
	1.35				24
	1.40				67
	1.45				62
	1.50				42
	1.55				81
6	1.60				

In summary, both the deterministic and stochastic equations have the same steady-state mean value. The stochastic approach indicates a finite extinction possibility, whereas the deterministic does not. In the deterministic case we obtain the smooth curve in Fig. 2.3 and in the stochastic case we have the discontinuous curve. The possibility of extinction is particularly important when N is small, that is, not much larger than unity, and therefore it would be important, in this case, to perform the stochastic analysis.

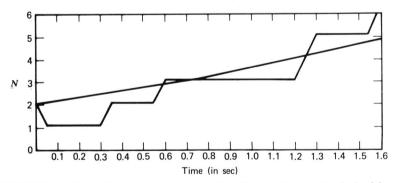

FIGURE 2.3. Population N as a function of time. The smooth curve is obtained from the deterministic equation and the broken curve is from the stochastic equations.

2.3
THE TWO-GROUP PROBLEM

In Section 2.2 we considered birth and death processes in a single group, and we shall now extend the analysis to two interacting systems. Several examples to be considered are the numbers in two different age groups, the relationship between oxygen and effluence concentrations in water, behavior of electrons and light in a laser, and interactions between a predator population and a prey population.

Consider two populations, $N_1(t)$ and $N_2(t)$. The rate of change of population $N_1(t)$ with time may be written as

$$\frac{dN_1}{dt} = (\text{number of type } N_1 \text{ injected or migrating}$$
$$\text{into the system per unit time})$$

$$+ (\text{number of births per unit time of type } N_1)$$

$$+ (\text{number of type } N_2 \text{ changing to type } N_1 \text{ per unit time})$$

$$- (\text{number of type } N_1 \text{ ejected or leaving per unit time})$$

$$- (\text{number of deaths per unit time of type } N_1)$$

$$- (\text{number of type } N_1 \text{ changing to type } N_2 \text{ per unit time}).$$

A similar expression may be written for (dN_2/dt). Each of the terms in the rate equations may be a function of N_1, N_2, and time. Frequently, the various factors contributing to change are expressed as the first few terms of a power series expansion. If, for example, the probability of a birth per unit time per individual is relatively independent of the number, then the number of births per unit time is linearly proportional to the number. If deaths result from encounters between the two groups, the first term in the power series expansion for this effect involves the product $N_1 N_2$, since if either population becomes zero there are no encounters. The lowest-order term expressing the rate of change from type N_1 to N_2 is proportional to N_1, and similarly, the rate of change from N_2 to N_1 is proportional to N_2.

Certain terms in the rate equations may be common to both (dN_1/dt) and (dN_2/dt). A rate of transfer from N_1 to N_2 will contribute the same absolute value to (dN_1/dt) as to (dN_2/dt), but with a negative sign in the former equation and a positive sign in the latter. If there is mutual annihilation, that is, there are simultaneous deaths, then there is a common negative term in both rate equations.

Two Age Groups

As a first example of the two-group system, let us consider a human population of age 0–10 years as one subgroup, and from 10+ years on as the other. The former population will be designated as N_1 and the latter, as N_2. Figure 2.4 illustrates the flows of numbers resulting from births, deaths, and aging. We shall assume that the probability of a birth or death per unit time per individual is independent of population and time, so that

$$\frac{dN_1}{dt} = a_2 N_2 - b_1 N_1 - R(t) \tag{2.22}$$

$$\frac{dN_2}{dt} = R(t) - b_2 N_2, \tag{2.23}$$

where a_1, b_1, and b_2 are constants. The first term on the right-hand side of (2.22) is the number of births per unit time, which is proportional to the reproductive population N_2 giving birth. The second term is the number of deaths per unit time, and the third term is the number per unit time that transfer from the first group to the second due to aging. Equations (2.22) and (2.23) contain three dependent variables N_1, N_2, and R so that an additional equation is required to solve for these variables. The transfer rate R is equal to the number per unit time reaching their 10th birthday:

$$R(t) = a_2 \left[N_2(t-10) \right] S, \tag{2.24}$$

where $N_2(t-10)$ is the number in the 10+ group 10 years previously, and S is the fraction alive at time t born 10 years before. The product $a_2 N_2(t-10)$ is the number born per unit time at time $t = -10$ years. [If the birth rate, a_2, is also a function of time, then $a_2(t-10)$ must be used in (2.24).]

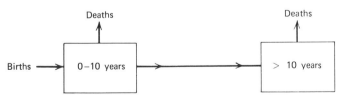

FIGURE 2.4.   Flows of populations between two age groups.

Survival rate S is less than unity because of deaths that occur between years 0 and 10. Due to deaths only,

$$\frac{dN_1}{dt} = -b_1 N_1, \tag{2.25}$$

from which one obtains

$$N_1(t) = N_1(t-10)e^{-10b_1}. \tag{2.26}$$

(If b_1 is a function of time, then e^{-10b_1} in (2.26) must be replaced by $\exp[-\int_{t-10}^{t} b_1(t)\,dt.]$) The survival rate is the ratio of the number alive at time t to the number alive at $t-10$:

$$S = \frac{N_1(t)}{N_1(t-10)} = e^{-10b_1}. \tag{2.27}$$

With (2.27) and (2.24) substituted into (2.23), we find

$$\frac{dN_2}{dt} = a_2' N_2(t-10) - b_2 N_2, \tag{2.28}$$

where $a_2' = a_2 e^{-10b_1}$. The solution to (2.28) is

$$N_2 = N_2(0)e^{pt}, \tag{2.29}$$

where p satisfies the transcendental equation

$$a_2' e^{-10p} - p - b_2 = 0. \tag{2.30}$$

In a human population, $|a_2' - b_2| \ll .1$, whereupon $p \cong \dfrac{a_2' - b_2}{1 + 10a_2'}$. Population N_1 is obtained from the simultaneous solution of (2.22), (2.23), and (2.29). If $a_2' < b_2$ then both N_1 and N_2 decay with time.

Water Pollution

An interesting example of two group interaction is the relationship between the concentration of oxygen dissolved in the water and the concentration of organic waste. The waste is degraded by bacteria, which produce a chemical reaction that involves the utilization of the dissolved oxygen. Waste concentration is often measured in terms of the biochemical oxygen demand (BOD), which is the amount of oxygen per unit volume of water required to degrade the waste. (The measurement unit for BOD might be milligrams of oxygen per liter of water.) The rate at which waste is degraded is proportional to the waste concentration, assuming that there is sufficient oxygen present in the water to allow the process to proceed. If we use the symbol L to represent BOD, then the rate of degradation is

$$\frac{dL}{dt} = -k_1 L, \tag{2.31}$$

where k_1 is known as the deoxygenation constant, and is typically measured in days^{-1}.

In the absence of waste, the oxygen concentration will be at an equilibrium value C_0. (C_0 is a known function of water temperature.) In the presence of waste, the actual concentration C will be below C_0, and we will define the oxygen concentration depletion D as

$$D = C_0 - C. \tag{2.32}$$

The variable D tends to increase with time due to the reduction of waste, and to decrease with time due to the absorption of oxygen at the surface of the water. (This latter process is known as *reaeration*.) Therefore we have

$$\frac{dD}{dt} = k_1 L - k_2 D, \tag{2.33}$$

where the first term on the right-hand side is the waste-reduction effect and the second term is reaeration. The constant k_2 is known as the reaeration constant measured in days^{-1}.

Equations (2.31) and (2.33) are the Streeter–Phelps Equations.[4] The simultaneous solution of these equations gives

$$D = \frac{k_1}{k_2 - k_1} L(0)(e^{-k_1 t} - e^{-k_2 t}) + D(0)e^{-k_2 t}, \tag{2.34}$$

where $L(0)$ is the BOD concentration at $t = 0$ and $D(0)$ is the depletion concentration at $t = 0$.

When there are a number of business activities distributed along a stream, each contributing some pollution to the water, a factor of interest is the maximum oxygen-depletion concentration that occurs.[5] If, for example, the oxygen concentration falls below a certain level, fish will not survive. The time variable t is related to the distance, x, along the stream by $x = vt$, where v is the velocity of the water. Therefore we may substitute

$t = (x/v)$ into (2.34) and obtain a curve of D as a function of x. Figure 2.5 is a plot of oxygen concentration as a function of distance. The maximum depletion D_M is found by setting the derivative of (2.34) equal to zero:

$$D_M = L(0)\frac{k_1}{k_2}\left[\frac{k_2}{k_1}\left(1 - \frac{D(0)(k_2 - k_1)}{L(0)k_1}\right)\right]^{k_1/(k_1 - k_2)}, \qquad (2.35)$$

where $L(0)$ is the BOD concentration in the stream at $x = 0$ and $D(0)$ is the initial depletion concentration, which is nonzero because of upstream pollutors. If D_M is constrained to be less than a specified value and $D(0)$, k_1, and k_2 are fixed, then the maximum amount of BOD that may be injected at $x = 0$ is determined by solving (2.35) for $L(0)$.

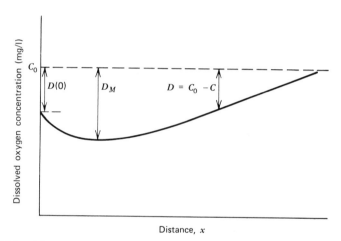

FIGURE 2.5. Dissolved oxygen concentration as a function of distance. Deficiency D is the difference between the concentration in the absence of waste C_0 and the actual concentration C. $D(0)$ is the depletion at $x = 0$, and D_M is the maximum depletion.

The Laser

The laser, a device for providing light amplification, is an example of a physical two-group interaction system.[6] This example is included to illustrate the strong similarity between continuity equations in very different fields. The concepts are the same, only the vocabularies differ. In particular, the laser and the predator–prey equations given in the following section are almost identical.

According to quantum theory, the energy of an electron in its orbit about a nucleus can assume only discrete values. Each energy value is associated with a different state, or condition, of the electron. Light amplification occurs at a frequency ω when the energy difference between two states of the electron equals $\hbar\omega$, where $\hbar$ is known as Planck's constant, and more electrons are in the higher energy state than in the lower energy state. If there are N_1 electrons per unit volume in the upper energy state and N_2 per unit volume in the lower energy state, then we want the population difference per unit volume N, where $N = N_1 - N_2$, to be positive.

Two factors affect the rate of change of N with respect to time. First, in the absence of light there is a rate of flow of electrons from the higher to the lower energy state toward an equilibrium value N^e. (N^e is the equilibrium population difference in the presence of some excitation, such as a flash lamp, that causes N_1 in equilibrium to exceed N_2 in equilibrium.) An upper-state electron may lose energy by collisions between particles or by the spontaneous emission of light. This flow proceeds at a rate proportional to the difference $(N - N^e)$. Secondly, the population difference is altered by the presence of light at frequency ω. The light tends to reduce N to zero, at a rate proportional to the product of N and the energy density, φ, of the light. Therefore we may write the equation

$$\frac{dN}{dt} = -\frac{(N - N^e)}{T} - K\varphi N. \qquad (2.36)$$

If $\varphi = 0$, (i.e., no light) then N varies as $e^{-t/T}$, so that T is the time constant associated with the flow of electrons from the upper to lower state in the absence of light. The second term on the right-hand side of (2.36) is the change induced by the presence of light, where K is a proportionality constant.

The rate of change of light energy density is also influenced by two effects. There is a natural decay that results from losses in the associated circuitry, and this loss rate is proportional to φ. Secondly, when the light induces an electron to change from an upper to a lower energy state, there is a corresponding increase in the energy density of the light. We may write

$$\frac{d\varphi}{dt} = -\frac{\varphi}{\tau} + \frac{K}{2}\varphi N, \qquad (2.37)$$

where τ is the time constant associated with the natural decay of φ. The factor of $\frac{1}{2}$ multiplies K in (2.37) because a single electron switch from an upper to lower state changes population difference N by 2, and produces a unit change in φ. Figure 2.6a is a plot of N as a function of time, Fig. 2.6b shows φ as a function of time, and Fig. 2.6c is the corresponding curve in the $\varphi - N$ plane.

Steady-state values for N and φ are obtained from (2.36) and (2.37) by setting the time derivatives equal to zero in these equations. These values, designated as N_0 and φ_0, are shown in Fig. 2.6. The arrow on the curve in Fig. 2.6c shows the direction of increasing time in the $\varphi - N$ plane. The light emitted by the laser is proportional to φ, so that the curve in Fig. 2.6b indicates the time dependence of the emission.

An important consideration is whether or not the steady-state solution to (2.36) and (2.37) is stable in the immediate vicinity of this solution. If, for example, we have reached steady state and then disturb the system slightly and the system does not return to the steady-state solution, this solution is unstable. The system settles at the steady state only if it is stable. From (2.36) and (2.37) we find that the steady-state values N_0 and φ_0 [obtained by setting $(d\varphi / dt)$ and (dN / dt) to zero] are

$$N_0 = \frac{2}{\tau K} \qquad \varphi_0 = \frac{\tau N^e}{2T} - \frac{1}{KT}. \tag{2.38}$$

To check stability near steady state, we perturb the system slightly from N_0 and φ_0 by letting

$$N = N_0[1 + \lambda f_1(t)], \tag{2.39}$$

and

$$\varphi = \varphi_0[1 + \lambda f_2(t)], \tag{2.40}$$

where $\lambda \ll 1$. To first-order terms in λ, the substitution of (2.39) and (2.40) into (2.36) and (2.37) yields

$$\dot{f}_1 + \frac{K\tau N^e}{2T} f_1 = \left[\frac{1}{T} - \frac{K\tau N^e}{2T} \right] f_2 \tag{2.41}$$

$$\dot{f}_2 = \frac{f_1}{\tau}. \tag{2.42}$$

If $f_2(t)$ is eliminated from (2.41) and (2.42), we obtain

$$\ddot{f}_1 + \frac{K\tau N^e}{2T} \dot{f}_1 = \frac{1}{T} \left[\frac{1}{\tau} - \frac{KN^e}{2} \right] f_1. \tag{2.43}$$

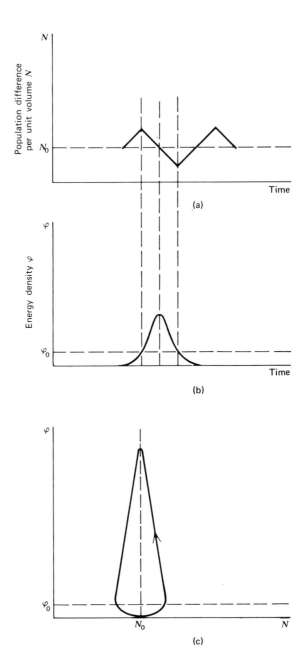

FIGURE 2.6. Population difference per unit volume N and energy density, φ, as functions of time, and in the φ-N plane. These curves were obtained from the solution to Eqs. (2.36) and (2.37).

The solutions to (2.43) are characterized by decaying exponentials if $(KN^e/2) > (1/\tau)$ and by growing exponentials if $(KN^e/2) < (1/\tau)$. Therefore, steady-state laser action is obtained only when

$$\frac{KN^e}{2} > \frac{1}{\tau}, \tag{2.44}$$

for only under this condition does a disturbance from the N_0, φ_0 equilibrium decay to zero.

In addition to the N_0, φ_0 solution to (2.36) and (2.37), another solution, corresponding to zero radiation, is

$$\begin{aligned}\varphi &= 0 \\ N &= N^e.\end{aligned} \tag{2.45}$$

This solution is stable when $(KN^e/2) < (1/\tau)$ and unstable when $(KN^e/2) > (1/\tau)$. The inequality expressed by (2.44) is known as the *threshold condition for oscillation*. If the laser medium is not excited sufficiently to have the equilibrium population difference in the presence of the excitation N^e large enough to satisfy (2.44), then laser radiation does not occur. In this case the $\varphi = 0$, $N = N^e$ solution is stable and there is no radiation.

Predator–Prey

A set of equations very similar to (2.36) and (2.37) for the laser were developed by Lotka and Volterra to describe interactions between predator and prey groups.[7] If the equilibrium population difference N^e is set equal to zero and T is negative in (2.36), then the laser and Lotka–Volterra equations are identical. Energy density φ is analogous to the predator in that it reduces the population difference, and N has the role of the prey. The interaction term in both examples is proportional to the product of the variables, since there are no encounters between predator and prey if either number is reduced to zero.

There are several reasons for the interest in modeling the growth and decay of animal populations. A breed may be harvested to provide food or other materials, or control of a particular animal group may be desired to protect crops or another species. Since the original modeling attempt by Lotka and Volterra there have been extensions of the model to account for variations in behavior with different age groups, to include spatial distributions in populations, to incorporate environmental fluctuations, and to provide improved correlation with experimental observations. We shall consider the original model, and the reader is referred to the literature for a discussion of these extensions.[8]

The equations for the predator–prey interaction are based upon the following assumptions:

1. In the absence of the predator, the number of prey grows at a rate proportional to its number.
2. In the absence of the prey, the number of predators diminishes at a rate proportional to its number.
3. The encounters between predator and prey cause the predator number to increase and the prey number to decrease, and due to encounters, populations change at a rate proportional to the product of the two populations.

This last assumption is based on the fact that the first term in the power series representation for encounters between two groups is proportional to the product of populations, since there are no encounters if either number is zero.

These assumptions lead to the following expressions:

$$\frac{dV}{dt} = b_1 V - b_2 A V \tag{2.46}$$

$$\frac{dA}{dt} = -b_3 A + b_4 A V, \tag{2.47}$$

where A is the number of predators; V is the number of prey; and b_1, b_2, b_3, and b_4 are positive constants. The steady-state solutions are obtained by setting the time derivatives to zero:

$$A_0 = b_1/b_2 \qquad V_0 = b_3/b_4,$$

where A_0 and V_0 are the steady-state populations.

It is convenient to normalize (2.46) and (2.47) to the steady-state values by defining the new variables a and v

$$a = A/A_0 \qquad v = V/V_0. \tag{2.48}$$

The substitution of (2.48) into (2.46) and (2.47) gives the normalized equations

$$\frac{da}{dt} = -b_3 a(1-v) \tag{2.49}$$

$$\frac{dv}{dt} = b_1 v(1-a). \tag{2.50}$$

These equations are nonlinear but are separable in a–v space. The ratio of (2.49) to (2.50) yields

$$\frac{da}{dv} = -\frac{b_3}{b_1} \frac{(1-v)a}{(1-a)v}, \tag{2.51}$$

which may be integrated to give

$$b_1[\ln a - a] + b_3[\ln v - v] = K, \tag{2.52}$$

where K is a constant. Different values for K give a family of curves, one of which is shown in Fig. 2.7. The constant K is specified if a and v are given at any instant of time.

The arrow on the curve in Fig. 2.7 shows the direction of increasing time. As time increases, a and v progress in a counterclockwise direction

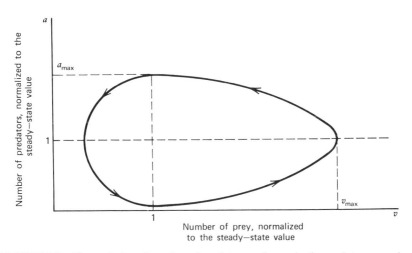

FIGURE 2.7. The variation of number of predators and prey in the predator–prey plane.

around the curve in a periodic fashion, with a time period determined by the amount of time required to traverse one complete loop. The variables a and v never reach the steady-state point located at $(1, 1)$ in the $a - v$ plane.

Suppose our policy is to increase a_{max}, the maximum number of predators. If we are at point P_1 on curve C_1 in Fig. 2.8, we may harvest predators to move to point P_2 on curve C_2. The maximum number of predators attained on curve C_2 exceeds the maximum attained on C_1, and so by harvesting predators at an appropriate point in the cycle it has been possible to increase the maximum. This result occurs because harvesting predators at point P_1 allows the number of prey to increase beyond the level attained without harvesting (v_{max} on C_2 is greater than v_{max} on C_1). Additional prey provide a larger food supply for the predators so that the peak number of predators increases.

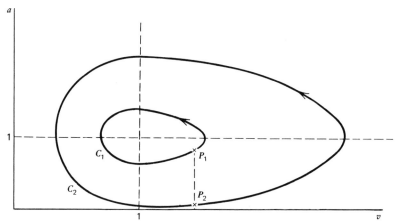

FIGURE 2.8. Harvesting predators at P_1 leads to an increase in the maximum number of predators.

In this section our rate equations have been considered to be deterministic. However, one could treat the equations, such as (2.46) and (2.47) stochastically, as was done in the previous section. Particularly for the case of small population numbers, where there is a finite probability of extinction, the stochastic analysis could lead to a result markedly different from that obtained by the deterministic approach.

The stability of the steady-state solution to (2.49) and (2.50), $a = v = 1$, may be examined by the same approach used to consider the stability of the steady-state solution to the laser equations. Namely, we allow a small perturbation from steady state by writing

$$a = 1 + \lambda f_1(t) \tag{2.53}$$

$$v = 1 + \lambda f_2(t), \tag{2.54}$$

where $\lambda \ll 1$. The substitution of (2.53) and (2.54) into (2.49) and (2.50) yields

$$\dot{f}_1 = b_3 f_2 \tag{2.55}$$

$$\dot{f}_2 = -b_1 f_1. \tag{2.56}$$

From (2.55) and (2.56) we may obtain an equation solely in terms of f_1 or f_2 of the form

$$\ddot{f}_1 + b_1 b_3 f_1 = 0. \tag{2.57}$$

The solution to (2.57) is

$$f_1 = \alpha \cos \sqrt{b_1 b_3}\ t + \beta \sin \sqrt{b_1 b_2}\ t, \tag{2.58}$$

where α and β are constants. Thus a small disturbance results in undamped sinusoidal oscillations about the steady-state solution. This is a *conditionally stable solution*, for a perturbation neither grows nor decays in amplitude with time.

2.4
MULTIPLE-GROUP INTERACTION

Chemical Reactions

In the previous section we considered interactions between two different groups, and in this section we shall consider many interacting species. An important example of this type of problem concerns chemical reactions, such as might occur in the production of air pollutants.[9] For example, suppose chemicals A_1 and A_2 react to produce A_3 and A_4:

$$A_1 + A_2 \rightarrow A_3 + A_4.$$

This is an encounter type of problem in that the reaction does not occur if either the concentration of A_1, c_1, or the concentration of A_2, c_2, becomes zero. Therefore we may write

$$\frac{dc_1}{dt} = \frac{dc_2}{dt} = -kc_1c_2, \tag{2.59}$$

where k is the proportionality constant. Each unit decrease in A_1 and A_2 results in a corresponding unit increase in A_3 and A_4, so that

$$\frac{dc_3}{dt} = \frac{dc_4}{dt} = kc_1c_2, \tag{2.60}$$

where c_3 and c_4 are the concentrations of A_3 and A_4, respectively.

If three chemicals interact:

$$A_1 + A_2 + A_3 \rightarrow \sum_{i=4}^{N} A_i$$

then

$$\frac{dc_1}{dt} = \frac{dc_2}{dt} = \frac{dc_3}{dt} = -kc_1c_2c_3, \tag{2.61}$$

where k is the proportionality constant for the three chemical process. If $A_2 = A_3$, we have that

$$A_1 + 2A_2 \rightarrow \sum_{i=3}^{N} A_i$$

and

$$\frac{dc_1}{dt} = -kc_1c_2^2. \tag{2.62}$$

Equation (2.62) is obtained from (2.61) by letting $c_2 = c_3$. The concentration of A_2 changes twice as rapidly as. the concentration of A_1, since two of type A_2 are lost for each of type A_1. Therefore

$$\frac{1}{2}\frac{dc_2}{dt} = \frac{dc_1}{dt} = -kc_1c_2^2. \tag{2.63}$$

In general, for

$$\sum_{i=1}^{l} n_i A_i \rightarrow \sum_{i=l+1}^{N} n_i A_i$$

then

$$\frac{1}{n_j}\frac{dc_j}{dt} = -k \prod_{i=1}^{l} c_i^{n_i} \qquad 1 \leqslant j \leqslant l$$

$$= +k \prod_{i=1}^{l} c_i^{n_i} \qquad l+1 \leqslant j,$$

$$(2.64)$$

where

$$\prod_{i=1}^{l} c_i^{n_i} = (c_1^{n_1})(c_2^{n_2})\cdots(c_l^{n_l}).$$

Equation (2.64) expresses the rate of change of concentration for an l-species encounter process.

Materials Balance

Another example of multiple-group interaction is the materials-balance description of the transfer of materials between the environment, industry and consumers.[10] Figure 2.9 illustrates the flow of materials between the environment, industry and consumers. The environment provides both natural products, such as fossil fuels, minerals, water, and oxygen, as well as agricultural and animal products. These resources are used by industry. (There is some direct use of environmental resources, e.g., water and air, by consumers, but this use has not been included in the model.) Industry provides products for consumers and returns residuals, such as the by-products of production, to the environment. Some consumer items may be recycled to be used again by industry, and the remainder eventually return to the environment as residuals. (Consumers include private purchasers, the federal government, foreign nations, and state and local governments.)

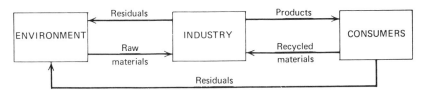

FIGURE 2.9.   Materials flow between the environment and sectors of the economy.

Any difference between input and output flows to a sector of the economy must result in a mass change within the sector. The variables are:

X_k = number of items of product type k produced per unit time.

Y_k = number of items of product type k obtained by consumers per unit time. (In general, $X_k \neq Y_k$, since some production items are used within the industrial sector rather than by consumers.)

M_i^E = material mass of type i in the environment

M_i^I = material mass of type i in the industrial sector

M_i^C = material mass of type i in the consumption sector

The material balance equations are:

$$\frac{dM_i^E}{dt} = - \sum_j A_{ij} X_j + \sum_j B_{ij} X_j + \sum_j C_{ij} M_j^C \qquad (2.65)$$

(raw mat-   (residual   (residual
erial mass   mass per   mass per
per unit   unit time   unit time
time)   from in-   from con-
                   dustry)   sumers)

$$\frac{dM_i^I}{dt} = \sum_j A_{ij} X_j - \sum_j B_{ij} X_j - \sum_{jj} D_{ij} Y_j + \sum_j E_{ij} M_j^C \qquad (2.66)$$

(raw mat-   (residual   (mass per   (recycled
erial mass   mass per   unit time   material
per unit   unit time   to consum-   mass per
time)   from in-   ers in the   unit time)
                   dustry)   form of pro-
                                      ducts)

$$\frac{dM_i^C}{dt} = \sum_j D_{ij} Y_j \ - \ \sum_j E_{ij} M_j^C \ - \ \sum_j C_{ij} M_j^C \qquad (2.67)$$

(mass per unit time to consumers in the form of products)	(recycled material mass per unit time)	(residual mass per unit time from consumers)

where

A_{ij} = mass of material type i required to produce one item of type j

B_{ij} = mass of residual of type i from industry generated per unit of item of type j

C_{ij} = mass of residual of type i from consumers generated per unit time per unit mass of type j material

D_{ij} = mass of material of type i used in a consumer product of type j

E_{ij} = mass of material of type i recycled per unit time per unit mass of type j material.

The above coefficients may be functions of time and the X and Y variables, although generally these coefficients are taken to be constant.

If a steady-state exists in which there is no change in material mass in each of three sectors, (2.65) and (2.66) may be written as

$$\sum_j A_{ij} X_j \ = \ \sum_j B_{ij} X_j + \sum_j C_{ij} M_j^C \qquad (2.68)$$

(raw material mass per unit time)	(total residual mass per unit time)

$$\sum_j A_{ij}X_j = \sum_j B_{ij}X_j + \sum_j D_{ij}Y_j - \sum_j E_{ij}M_j^C$$

(raw mat-erial mass per unit time)	(residual mass per unit time from indus-try)	(mass per unit time to consum-ers in the form of pro-ducts)	(recycled material mass per unit time)

(2.69)

The equation obtained from (2.67) is not independent, for it may be derived by subtracting (2.68) from (2.69). From (2.68) we see that in the steady-state the total residual mass per unit time equals the mass of raw materials removed from the environment per unit time. An application of the above equations is to indicate the various options for reducing the raw material mass flow and thereby reducing the rate of residual generation:

• Consumption Y_j may be altered. Since production, X_j, depends on consumption, this variable will also be altered.
• Technology and processing may be improved so that residual generation is reduced; that is, reduce B_{ij} and C_{ij}.
• Recycling may be increased; that is, increase E_{ij}.
• Technology and processing may be improved so that the materials required to meet final demand is less; that is, reduce D_{ij}.

2.5
SEVERAL INDEPENDENT VARIABLES

The previous examples presented in this chapter considered the number of members of a group only as a function of time. However we may wish to express these numbers in terms of additional variables, such as the age or location of a member of the group.

Population as a Function of Age and Time

Suppose we are interested in the number in an age increment da at time t, where

$$N(a,t)da = \text{number in the age increment from } a \text{ to } a + da \text{ at time } t. \qquad (2.70)$$

If we consider a deathrate that is independent of concentration but may be a function of age and time, we have that

$$\frac{dN(a,t)}{dt} = -d(a,t)N(a,t) \qquad a>0, \qquad (2.71)$$

where $d(a,t)$ is the probability of a death per unit time per individual. Equation (2.71) does not hold for $a=0$ since $N(0,t)$ is determined by the birthrate rather than the deathrate.

Since $N(a,t)$ is a function of two variables, the time derivative is equal to

$$\frac{dN(a,t)}{dt} = \frac{\partial N(a,t)}{\partial t} + \frac{\partial N(a,t)}{\partial a}\frac{da}{dt}. \qquad (2.72)$$

To evaluate (2.72), we note that individuals age according to the relationship

$$a = a_0 + t, \qquad (2.73)$$

where a_0 is the age at $t=0$. Therefore (2.72) becomes

$$\frac{dN(a,t)}{dt} = \frac{\partial N(a,t)}{\partial t} + \frac{\partial N(a,t)}{\partial a}. \qquad (2.74)$$

The combination of (2.71) and (2.74) yields

$$\frac{\partial N(a,t)}{\partial t} + \frac{\partial N(a,t)}{\partial a} = -d(a,t)N(a,t) \qquad a>0. \qquad (2.75)$$

To solve for $N(a,t)$ from (2.75) it is necessary to determine the boundary conditions $N(a,0)$ and $N(0,t)$. The former condition may be specified by measuring the population distribution at $t=0$, and the latter condition is evaluated from the birthrate. Specifically, the number of individuals in the age interval from zero to da, $N(0,t)da$, equals the number born per unit time multiplied by the time dt required for an individual to age by an amount da. From (2.73) we have that $da/dt=1$, so that $N(0,t)$ equals the number born per unit time. Therefore

$$N(0,t) = \int_0^\infty b(a,t)N(a,t)da, \qquad (2.76)$$

where $b(a,t)$ is the probability of a birth per unit time per individual. Since $N(a,t)da$ is the number of individuals in the age increment da, the integral in (2.76) is the number of births per unit time. With the rates $b(a,t)$ and $d(a,t)$ given and $N(a,0)$ specified, the population distribution as a function of time can be calculated from (2.75) and (2.76).

If a steady-state condition is reached where time is not an independent variable, (2.75) becomes

$$\frac{\partial N(a)}{\partial a} = -d(a)N(a),\qquad(2.77)$$

which has the solution

$$N(a) = N(0)\exp\left[-\int_0^a d(x)dx\right].\qquad(2.78)$$

We see that in the steady state, the population per unit age increment (i.e., the population distribution with age) is a function of the deathrate only. This is a reasonable result, for only deaths can alter the population from one age increment to another. From (2.76) we have that

$$N(0) = \int_0^\infty b(a)N(a)da,\qquad(2.79)$$

which, when combined with (2.78), yields

$$1 = \int_0^\infty da\,b(a)\,\exp\left[-\int_0^a d(x)dx\right].\qquad(2.80)$$

Equation (2.80) specifies a relationship between birth and death rates that must hold if a steady-state condition is to be achieved.

Variable Dependence on Space and Time; Air Pollution Continuity Equation

An important example of variable dependence on space and time concerns the concentration of air pollutants.[11] If c_i is the concentration of the ith pollutant (i.e., number of particles per unit volume), then the continuity equation is

$$\frac{\partial c_i}{\partial t} =$$ (rate of increase of concentration resulting from particle motion due to the movement of the air)

+ (rate of increase resulting from motion of the particles relative to the air)

+ (rate of increase due to a pollutant source)

− (rate of decrease due to a sink).

(2.81)

We shall now proceed to derive the mathematical formulation for (2.81). The current density $\mathbf{J}_i$, where

$\mathbf{J}_i$ = number of particles of type i per unit area per
unit time crossing a surface normal to the direction
of flow

is related to the velocity $\mathbf{v}_i$ of the ith particle by the equation[12]

$$\mathbf{J}_i = c_i \mathbf{v}_i. \tag{2.82}$$

In a volume element where no particles are generated or eliminated, we have an additional relationship based on conservation of particles given by[13]

$$\nabla \cdot \mathbf{J}_i = -\frac{\partial c_i}{\partial t} \tag{2.83}$$

where, in rectangular coordinates,

$$\nabla \cdot \mathbf{J}_i = \frac{\partial J_{ix}}{\partial x} + \frac{\partial J_{iy}}{\partial y} + \frac{\partial J_{iz}}{\partial z}, \tag{2.84}$$

and J_{ix}, J_{iy}, and J_{iz} are the x, y, and z components respectively of current density. If we substitute (2.82) into (2.83), the particle conservation equation becomes

$$-\frac{\partial c_i}{\partial t} = \nabla \cdot (c_i \mathbf{v}_i) = c_i \nabla \cdot \mathbf{v}_i + (\mathbf{v}_i \cdot \nabla)c_i, \tag{2.85}$$

where, in rectangular coordinates,

$$(\mathbf{v}_i \cdot \nabla)c_i = \left(v_{ix}\frac{\partial}{\partial x} + v_{iy}\frac{\partial}{\partial y} + v_{iz}\frac{\partial}{\partial z} \right)c_i.$$

For a medium that is incompressible, that is, where the particle density is not a function of time or space, from (2.85) we have

$$-\frac{\partial c_m}{\partial t} = 0 = c_m \nabla \cdot \mathbf{v}_m + (\mathbf{v}_m \cdot \nabla)c_m, \tag{2.86}$$

where c_m and $\mathbf{v}_m$ are the particle concentration and velocity for the medium. Since c_m does not depend on coordinate location, $(\mathbf{v}_m \cdot \nabla)c_m = 0$ and therefore (2.86) yields

$$\nabla \cdot \mathbf{v}_m = 0. \tag{2.87}$$

For the case of air pollution it is reasonable to assume that the medium (i.e., the air) is incompressible. With the substitution of (2.87) into (2.85), we obtain

$$\frac{\partial c_i}{\partial t} + (\mathbf{v}_m \cdot \nabla)c_i = -\nabla \cdot (c_i \boldsymbol{\nu}_i) = -\nabla \cdot \mathbf{j}_i, \qquad (2.88)$$

where $\boldsymbol{\nu}_i$ is the pollutant velocity relative to the medium ($\boldsymbol{\nu}_i = \mathbf{v}_i - \mathbf{v}_m$) and $\mathbf{j}_i$ is the pollutant current density relative to the medium: $\mathbf{j}_i = c_i \boldsymbol{\nu}_i$.

Equation (2.88) expresses the conservation of particles for a pollutant in an incompressible medium. The first term on the left-hand side is the rate of change of concentration, the second term is the rate of change of concentration that results from motion of the medium, and the right-hand term is the rate of change of concentration that results from particle motion relative to the medium.

If particle sources or sinks are present at any point, then additional terms must be added to (2.88). A source might be a smokestack, a sink might be the ground (as for particulates), and a chemical reaction might be either a source or sink for a given pollutant. With

$$S_i = \text{rate of concentration increase of the } i^{\text{th}}$$
pollutant due to a source

$$D_i = \text{rate of concentration decrease due to a sink}$$

(2.88) is modified to become

$$\frac{\partial c_i}{\partial t} = -(\mathbf{v}_m \cdot \nabla)c_i - \nabla \cdot \mathbf{j}_i + S_i - D_i. \qquad (2.89)$$

Current density relative to the medium $\mathbf{j}_i$ may result from a velocity $\boldsymbol{\nu}_i$ imposed by an auto exhaust, a smokestack, or some other source, or it may result from *diffusion* of the pollutant. If a pollutant is initially localized at one point in space, due to the random motion of the particles the contaminant will diffuse throughout the medium until a uniform distribution is reached. Figure 2.10 is a one-dimensional illustration of diffusion. Initially, at $t = t_0$, all of the particles are located at $x = 0$. Because of the symmetry of the system, there is equal likelihood that the random motion of the particle will take it in either the plus or minus directions. Therefore at t_1 approximately one half of the particles are displaced to the right and

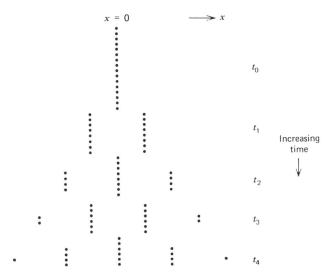

FIGURE 2.10.   The spread of particles resulting from diffusion.

the other half, to the left. As time increases the particles spread symmetrically about $x = 0$. This flow of particles corresponds to a current density proportional to the change of concentration with respect to distance: In one dimension

$$j_{ix} = k_x \frac{\partial c_i}{\partial x},\qquad (2.90)$$

where k_x is known as the diffusivity constant in the x-direction. When the concentration is uniform (i.e., $\partial c_i / \partial x = 0$), there is no further diffusion current. For the case of air pollutants, diffusion effects are modified by the gravitational force, by air resistance, and by buoyancy.

Boundary Conditions

If sources, sinks, and medium velocity can be characterized mathematically, then the concentration c_1 is determined from (2.89) upon specification of the boundary conditions on the region. The earth may act as a sink for some contaminants (as particulates) or it may act as a containing wall. As a containing wall the boundary condition is that the current density in the normal direction to the surface is zero. This means that the diffusivity constant in the normal direction and the velocity in the normal direction are both zero.

Another containing wall may be established in a plane parallel to the earth's surface if certain atmospheric temperature conditions exist. Figure 2.11 illustrates a possible atmospheric temperature dependence as a function of the height z above the earth's surface. Lapse rate L is defined as

$$L = - \frac{\partial T}{\partial z}, \tag{2.91}$$

where T is the temperature.

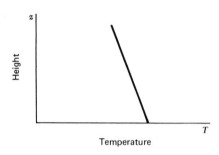

FIGURE 2.11. Atmospheric temperature dependence upon height above the earth's surface.

When a column of air rises it expands because of decreasing pressure, and this expansion results in a reduction in temperature. If the temperature of the surrounding air is decreasing at the same rate as the rising column, there is no exchange of heat between the column and the atmosphere. In this case the process is adiabatic, that is, there is no exchange of energy between the rising column and its surroundings. The lapse rate for which this condition holds is known as the *adiabatic lapse rate*, and is approximately 5.5 °F per 1000 ft of height.

If the lapse rate is adiabatic, then any given volume of air is at the same temperature and pressure as its surroundings so that there are no net forces exerted on the volume. This is a condition of unstable equilibrium, since a small applied force will cause an air volume to move essentially without restraint. If the lapse rate is faster than adiabatic surrounding (i.e., the air cools off faster with increasing height than does a rising volume) as shown in Fig. 2.12a, then a rising air volume is warmer and less dense than its surroundings so that the surrounding air exerts a net upward force (buoyancy) on the volume and it continues to rise. Similarly, a falling volume continues to fall. Therefore, a lapse rate that is faster than adiabatic results in an unstable atmosphere.

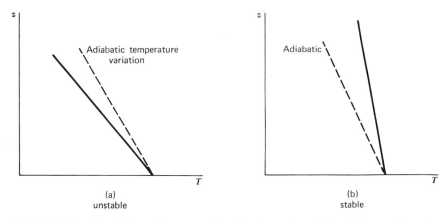

FIGURE 2.12. Two different lapse rates. The first, faster than adiabatic, is unstable, and the second is stable.

When the lapse rate is less than adiabatic, as shown in Fig. 2.12*b*, a rising air volume is at a lower temperature and more dense than its surroundings, so that the net force exerted on the volume is downward. Similarly, a falling volume experiences an upward force. This is a stable atmosphere, since the displacement of an air volume results in a restoring force.

Figure 2.13*a* illustrates a possible temperature dependence that might occur in the morning, where the upper atmosphere, with a shorter thermal time constant than the earth, is warmed by the rising sun. Since the atmosphere is stable above height *h*, termed the *inversion height*, there would be little penetration of pollutants above this level from the two smokestacks shown in the figure. Figure 2.13*b* might represent a midday temperature variation, after the earth has been warmed. In this case, the smokestack plumes are not constrained within an upper limit.

At night, in a rural setting where the ground is cold, the temperature dependence might be as shown in Fig. 2.13*c*. The lapse rate is more positive than the adiabatic rate, and so the atmosphere is stable for all heights and the smokestack plumes do not spread appreciably. At night, in an urban setting where the ground remains warm because of the buildings, the temperature variation might be shown in Fig. 2.13*d*. The atmosphere is in unstable equilibrium at the lower stack height. Therefore, the lower plume spreads more rapidly than the upper plume, with little penetration of the lower plume above height *h*. At the inversion height the boundary condition is the same as for a containing wall.

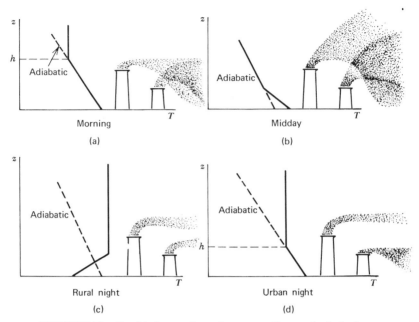

FIGURE 2.13. Possible lapse rates and corresponding smokestack plumes.

The boundary conditions on planes normal to the earth's surface may be determined by a mountain range that could either act as a sink or containing wall, similar to the earth's surface. If no mountain ranges or other obstacles are present, these boundaries may be chosen sufficiently far from the point at which the concentration is being calculated, so that either an absorbing or containing boundary may be assumed without appreciably altering the result.

A rather simple application of (2.89) may be used to obtain a rough estimate of the pollution concentration in a region. In this approximation, the box model shown in Fig. 2.14, it is assumed that the system is in steady state with no spatial dependence within the box on the x and z coordinates; that there is no particle motion relative to the medium; and that no sinks are within the box. This means that

$$\frac{\partial c_i}{\partial t} = j_i = D_i = 0,$$

whereupon (2.89) becomes

$$v_m \frac{\partial c_i}{\partial y} = S_i. \tag{2.92}$$

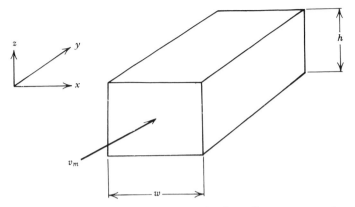

FIGURE 2.14.   The box model for estimating pollutant concentration.

If we define q_i as the amount of pollutant generated per unit time per unit length in the y-direction, then S_i, the amount generated per unit time per unit volume, is

$$S_i = \frac{q_i}{wh},$$ (2.93)

where w is the box width and h is the inversion height. Width w may be determined by natural features (e.g., mountain ranges) or perhaps by the dimensions of an urban area. From (2.92) and (2.93) we find that

$$c_i(y) = \frac{1}{v_m wh} \int_0^y q_i \, dy$$

$$= \frac{Q_i(y)}{v_m wh}$$ (2.94)

where $Q_i(y) = \int_0^y q_i \, dy$ is the amount of pollutant generated per unit time upwind from point y. Equation (2.94) may be used to obtain an estimate of the concentration without performing a detailed solution of (2.89).

2.6
AN URBAN GROWTH MODEL

An interesting example of the application of conservation equations to an urban growth model is presented by Forrester.[14] It is assumed in this model that events in the urban region do not appreciably alter the environment outside the city. It is taken that the number of people flowing into and out of the city does not significantly change the number living in outside areas, and that the rate of flow is determined solely by conditions within the city. In some cases this assumption is questionable, for instance, when the flow of people is largely to and from the immediate suburban region, where the population might be less than the urban population. The model also assumes certain behavior patterns for different population groups and certain structural interactions between urban subsystems. Most of these assumptions have been questioned.[15]

However regardless of the validity of these assumptions, the approach is interesting and may form the basis for improved future urban models. One advantage of this model is that it is specific in its assumptions and approach, so that it presents a concrete groundwork for identifying avenues for building better models.

Forrester's urban system is structured on the basis of three subsystems: (a) business, (b) housing, and (c) population. Each of these three is, in turn, subdivided into three elements, as shown in Fig. 2.15. Rate equations may be written for each of the nine subsystems shown in this figure. For example, consider the rate of change with time of the number of under-employed U,

$$\frac{\partial U}{\partial t} = (\text{number of laborers becoming underemployed per unit time})$$

$+ (\text{number of underemployed arrivals into the urban area per unit time})$

$+ (\text{number of births per unit time into underemployed})$

$- (\text{number of underemployed becoming laborers per unit time})$

$- (\text{number of underemployed departures per unit time})$

$- (\text{number of underemployed deaths per unit time})$ \hfill (2.95)

It has been assumed that there is no direct flow from the managerial–professional group to the underemployed or vice versa. Forrester writes his rate equations in difference rather than differential form, but the concepts are the same for both cases.

Forrester hypothesizes expressions for each of the flow rates on the right-hand side of (2.95). For example, the number of underemployed arrivals per unit time UA is taken to be proportional to the product of the sum of the underemployed U and laborers L, and the perceived attractive-

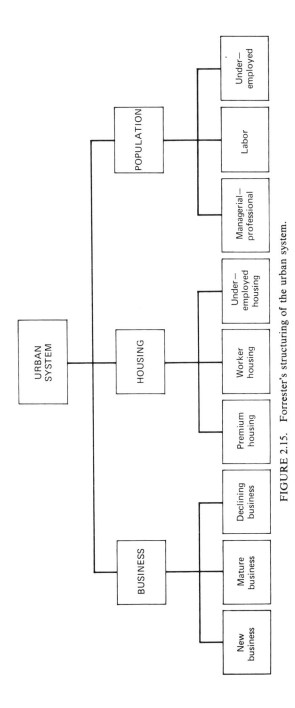

FIGURE 2.15. Forrester's structuring of the urban system.

ness PA of the urban area for the underemployed

$$UA = K(U+L)(PA), \qquad (2.96)$$

where K is a proportionality constant. Constant K is the ratio of under-employed arriving per unit time to $(U+L)$ when PA is unity.

Perceived attractiveness is related to the actual attractiveness of the area A by a time delay τ. That is, if some action is taken in the urban area to make it more attractive to the underemployed, it requires a time lapse of τ for the underemployed to perceive, understand, and react to this change. A differential relationship between PA and A yielding a delayed response is

$$\frac{d(PA)}{dt} + \frac{(PA)-A}{\tau} = 0, \qquad (2.97)$$

which has a solution

$$(PA) = A + \left[(PA)_0 - A\right]e^{-t/\tau}, \qquad (2.98)$$

where $(PA)_0$ is the perceived attraction at $t=0$. If, for $t<0$, the system is in equilibrium so that $(PA)=A$, and A is suddenly changed, (PA) approaches the new value of A with a time constant τ. The dependence of (PA) on time is shown in Fig. 2.16. As seen from this figure, (PA) approaches the new value of A with a time lag approximately equal to τ.

$$(PA)_{t+\Delta} = (PA)_t - \frac{\Delta}{\tau}\left[(PA)_t - A_t\right], \qquad (2.99)$$

where Δ is the time interval between calculations of (PA). In the limit as $\Delta \to 0$, (2.99) reverts to (2.97). Forrester assumed a value for τ equal to 20 years.

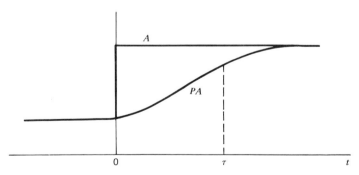

FIGURE 2.16.   The time lag between perceived attractiveness and attractiveness as given by Eq. (2.97).

The attractiveness of an urban area to the underemployed, A, is assumed to be a factor of five functions, representing the responses of the underemployed to different conditions. These factors, F_1 through F_5, are:

F_1: Representing the mobility of underemployed to labor group. Namely, how often is there a transition from underemployed to labor? An area is presumed to be more attractive if a larger fraction makes this transition in a given time period.

F_2: Representing the availability of housing for the underemployed. The more housing available, the more attractive is the area.

F_3: Representing the per capita public expenditure of funds. A city is more attractive if more is spent on public services.

F_4: Representing job availability. Attractiveness increases when more jobs are available to the underemployed.

F_5: Representing housing construction. The more houses constructed per unit time, the more desirable is the area.

Forrester adds a sixth factor F_6, which is not dependent upon any condition of the urban region but is included to allow A to vary independently of any other parameter in the system. This can be considered a sensitivity factor, which may be used to determine the sensitivity of the state variables to fluctuations in A. Attractiveness A is written as

$$A = \prod_{i=1}^{6} F_i.$$

The value for each factor F_1 through F_5 is obtained from a table relating the factor to the condition it represents. For example, F_1 is obtained from a table relating F_1 to the percentage of underemployed that move to the labor group per year. Table 2.2 enumerates this assumed dependence.

Table 2.2
Relationship Between F_1 and the Percentage From
Underemployed to Labor per Year

Percent from underemployed to labor per year	F_1
0	.3
2.5	.7
5.0	1.0
7.5	1.2
10.0	1.3
12.5	1.4
15.0	1.5

The value for the condition determining each of the factors is calculated as a function of the state variables and other parameters of the system. For example, the percentage moving from underemployed to labor each year is assumed to be a function of the number of new businesses, the number of mature businesses, the number of declining businesses, the number of laborers, the number of underemployed, and the tax per capita.

Thus far we have discussed the manner in which one of the flow rates (number of underemployed arrivals per unit time) on the right-hand side of (2.95) is specified. Each of the other flow rates in (2.95) are determined as follows:

1. Number of laborers becoming underemployed per unit time is taken to depend on the ratio of the number of laborers to the number of jobs for laborers. If this ratio is unity, a dropout rate of 3% per year is assumed, and the rate increases with an increasing ratio.
2. The difference between the birth and deathrates per year for underemployed is taken to be .015.
3. As stated in the previous paragraph, rate of flow from underemployed to labor is assumed to depend on a number of factors, including tax per capita, the number of new businesses, the number of laborers, and so on.
4. The underemployed departure rate is assumed to depend inversely on the factors that determine the underemployed arrival rate.

Finally, boundary conditions existing at the starting time are specified for the nine state variables shown in Fig. 2.15. Figure 2.17 illustrates some of the curves obtained from this model over a 250-year time span, using

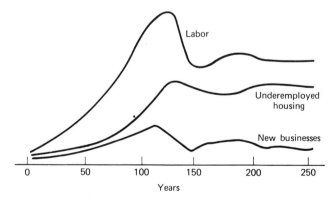

FIGURE 2.17. The variation of labor, underemployed housing, and new businesses in an urban area.

computer simulation. For the first one hundred years or so there is a rapid growth in all the state variables, for the next 100 years there are decreasing oscillations toward steady state until steady state is reached.

An important function of an urban growth model is to determine the impacts of the imposition of alternative policies. We note, for example, that the time constant for the variables in Fig. 2.17 is 50–100 years, so that the introduction of new policies would have a time lag of at least several decades. Also, if policies fluctuate in a time period short compared to the urban time constant, one would not expect significant changes to occur. This would suggest that the time span for urban planning should cover several decades.

Figure 2.18 illustrates the change in the variables in Fig. 2.17 once steady state has been reached and a program of low-cost housing is initiated. Each year, low-cost housing is provided for 2.5% of the under-employed, and therefore, the underemployed housing rises. However as a result of housing construction, the tax rate increases and the amount of available land is reduced. This renders the area less attractive to industry and so the number of new businesses decreases. Labor population drops because fewer jobs are available, and the lack of labor is an additional depressant on the construction of new business. Thus there is a significant reduction in both labor population and number of new businesses.

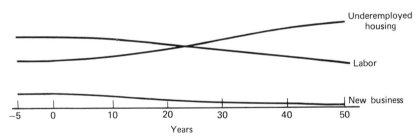

FIGURE 2.18. The change in variables with the introduction of a low-cost housing program.

REFERENCES

1. K. E. F. Watt, *Ecology and Resource Management*, McGraw-Hill, New York (1968), p. 292.
2. J. G. Kemeny and J. L. Snell, *Finite Markov Chains*, Van Nostrand, Princeton, N. J. (1959).
3. H. Chernoff and L. E. Moses, *Elementary Decision Theory*, 5th print., Wiley, New York (1967), p. 305.

4. H. W. Streeter and E. B. Phelps, *A Study of the Pollution and Natural Purification of the Ohio River*, U. S. Public Health Bull. 146 (Feb. 1925).

5. J. C. Liebman and W. R. Lynn, "The Optimum Allocation of Stream Dissolved Oxygen," *Water Resour. Res.* **2** (3), 581–591 (1966).

6. R. H. Pantell and H. E. Puthoff, *Fundamentals of Quantum Electronics*, Wiley, New York (1969), Ch. 4.

7. A. J. Lotka, *Elements of Physical Biology*, Williams & Wilkins, Baltimore (1925).

8. E. C. Pielou, *An Introduction to Mathematical Ecology*, Wiley, New York (1969), Ch. 6.

9. E. R. Stephens, "Chemistry of Atmospheric Oxidants," *J. of the Air Pollut. Control Assoc.* **19** (3), 181–185 (1969).

10. A. V. Kneese, R. U. Ayres, R. C. d'Arge, *Economics and the Environment, Materials Balance Approach*, Resources for the Future, Inc., Johns Hopkins Press, Baltimore (1970).

11. H. A. Panofsky, "Air Pollution Meteorology," *Amer. Sci.* **57** (2), 269–285 (1969).

12. See, for example, R. Plonsey and R. E. Collin, *Principles and Applications of Electromagnetic Fields*, McGraw-Hill, New York (1961), pp. 188–189.

13. Ibid., pp. 169–170.

14. J. W. Forrester, *Urban Dynamics*, M.I.T. Press, Cambridge, Mass. (1969), pp. 12–37.

15. A number of papers critiquing Forrester's model are included in IEEE Trans. on Systems, Man and Cybernetics, *SMC*-2, 2 (Apr. 1972).

PROBLEMS

2.1 A group population N has a probability of a birth per unit time given by

$$(.2 + .01N) \text{ s}^{-1},$$

and a probability of a death per unit time of

$$.02N \text{ s}^{-1}.$$

For an initial population $N = 5$, plot the solution to the deterministic equation. Calculate the steady-state solution. Using the table of random numbers given on page 92, plot 300 steps for the population as a function of time taking a time increment of .3 s.

2.2 If a stream has a rearation constant equal to .4 d^{-1} and a deoxygenation constant of .25 d^{-1}, what is the maximum allowable BOD concentration at a pollution site to avoid an oxygen deficiency in excess of 5 mg/l anywhere in the stream? Assume no upstream polluters.

2.3 Write a set of rate equations for population N_1, N_2, and N_3 where:

- In the absence of 2 and 3, group 1 grows at a rate proportional to its population.
- In the presence of group 3, group 1 decays because of encounters and group 3 grows.
- Group 2 grows because of encounters and consumption of members of group 3, and group 1 grows because encounters and consumption of group 2 members.

• In the absence of 1 and 3, group 2 grows because of a larger birthrate than deathrate but decays because of encounters between members of the group.
• Without 1 and 2, group 3 has achieved zero population growth. (Specify the signs of the coefficients that appear in your equations.)

2.4 A predator–prey ecosystem is in equilibrium with 100,000 predator and 75,000 prey. Suppose you wish to harvest the predator for food, and you want to obtain a larger supply. You decide to harvest 50,000 predator immediately without altering the prey population. Some time later you find that the predator and prey populations have grown to 80,000 and 112,500 respectively. What is the maximum number of prey you can expect to obtain? (Assume that the Lotka-Volterra equations apply.) If you had desired to obtain 10^6 predators in the shortest possible time, what should you have done to the equilibrium populations?

2.5 Consider the following flow diagram of university educated students. Define the necessary populations and write the appropriate difference equations for a one-year time increment. Assume that the rates of dropout, death and retirement are proportional to the respective populations. Is this a reasonable assumption? Have any flows been omitted from the diagram? If so, indicate what they are and whether or not you believe they are significant.

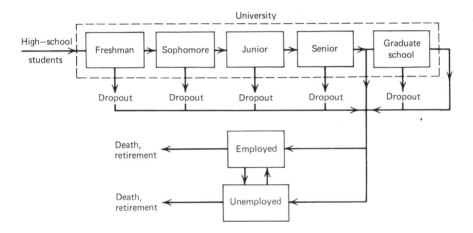

2.6 Derive a general expression for the concentration c_i of an air pollutant for zero wind velocity ($v_m = 0$), where S_i and D_i are also zero. (i.e., pollutant flow results primarilary from diffusion.)

2.7 Look up the most recent death rate figures in the United States as a function of age $d(a)$. If this death rate remains constant, compute the population distribution for a steady-state condition. What will be the percentage of people aged over 65 years and under 20 years? (See *Life Insurance Fact Book*, Institute of Life Insurance, New York.)

Table of Random Digits

03 47 43 73 86	36 96 47 36 61	46 98 63 71 62	33 26 16 80 45	60 11 14 10 05
97 74 24 67 62	42 81 14 57 20	42 53 32 37 32	27 07 36 07 51	24 51 79 89 73
16 76 62 27 66	56 50 26 71 07	32 90 79 78 53	13 55 38 58 59	88 97 54 14 10
12 56 85 99 26	96 96 68 27 31	05 03 72 93 15	5712 10 14 21	88 26 49 81 76
55 59 26 35 64	38 54 82 46 22	31 62 43 09 90	06 18 44 32 53	23 83 01 30 30
16 22 77 94 39	49 54 43 54 82	17 37 93 23 78	87 35 20 96 43	84 26 34 91 64
84 42 17 53 31	57 24 55 06 88	77 04 74 47 67	21 76 33 50 25	83 92 12 06 76
63 01 63 78 59	16 95 55 67 19	98 10 50 71 75	12 86 73 58 07	44 39 52 38 79
33 21 12 34 29	78 64 56 07 82	52 42 07 44 38	15 51 00 13 42	99 66 02 79 54
57 60 86 32 44	09 47 27 96 54	49 17 46 09 62	90 52 84 77 27	08 02 73 43 28
18 18 07 92 46	44 17 16 58 09	79 83 86 19 62	06 76 50 03 10	55 23 64 05 05
26 62 38 97 75	84 16 07 44 99	83 11 46 32 24	20 14 85 88 45	10 93 72 88 71
23 42 40 64 74	82 97 77 77 81	07 45 32 14 08	32 98 94 07 72	93 85 79 10 75
52 36 28 19 95	50 92 26 11 97	00 56 76 31 38	80 22 02 53 53	86 60 42 04 53
37 85 94 35 12	83 39 50 08 30	42 34 07 96 88	54 42 06 87 98	35 85 29 48 39
70 29 17 12 13	40 33 20 38 26	13 89 51 03 74	17 76 37 13 04	07 74 21 19 30
56 62 18 37 35	96 83 50 87 75	97 12 25 93 47	70 33 24 03 54	97 77 46 44 80
99 49 57 22 77	88 42 95 45 72	16 64 36 16 00	04 43 18 66 79	94 77 24 21 90
16 08 15 04 72	33 27 14 34 09	45 59 34 68 49	12 72 07 34 45	99 27 72 95 14
31 16 93 32 43	50 27 89 87 19	20 15 37 00 49	52 85 66 60 44	36 68 88 11 80
68 34 30 13 70	55 74 30 77 40	44 22 78 84 26	04 33 46 09 52	68 07 97 06 57
74 57 25 65 76	59 29 97 68 60	71 91 38 67 54	13 58 18 24 76	15 54 55 95 52
27 42 37 86 53	48 55 90 65 72	96 57 69 36 10	96 46 92 42 45	97 60 49 04 91
00 39 68 29 61	66 37 32 20 30	77 84 57 03 29	10 45 65 04 26	11 04 96 67 24
29 94 98 94 24	68 49 69 10 82	53 75 91 93 30	34 25 20 57 27	40 48 73 51 92
16 90 82 66 59	83 62 64 11 12	67 19 00 71 74	60 47 21 29 68	02 02 37 03 31
11 27 94 75 06	06 09 19 74 66	02 94 37 34 02	76 70 90 30 86	38 45 94 30 38
35 24 10 16 20	33 32 51 26 38	79 78 45 04 91	16 92 53 56 16	02 75 50 95 98
38 23 16 86 38	42 38 97 01 50	87 75 66 81 41	40 01 74 91 62	48 51 84 08 32
31 96 25 91 47	96 44 33 49 13	34 86 82 53 91	00 52 43 48 85	27 55 26 89 62
66 67 40 67 14	64 05 71 95 86	11 05 65 09 68	76 83 20 37 90	57 16 00 11 66
14 90 84 45 11	75 73 88 05 90	52 27 41 14 86	22 98 12 22 08	07 52 74 95 80
68 05 51 18 00	33 96 02 75 19	07 60 62 93 55	59 33 82 43 90	49 37 38 44 59
20 46 78 73 90	97 51 40 14 02	04 02 33 31 08	39 54 16 49 36	47 95 93 13 30
64 19 58 97 79	15 06 15 93 20	01 90 10 75 06	40 78 78 89 62	02 67 74 17 33
05 26 93 70 60	22 35 85 15 13	92 03 51 59 77	59 56 78 06 83	52 91 05 70 74
07 97 10 88 23	09 98 42 99 64	61 71 62 99 15	06 51 29 16 93	58 05 77 09 51
68 71 86 85 85	54 87 66 47 54	73 32 08 11 12	44 95 92 63 16	29 56 24 29 48
26 99 61 65 53	58 37 78 80 70	42 10 50 67 42	32 17 55 85 74	94 44 67 16 94
14 65 52 68 75	87 59 36 22 41	26 78 63 06 55	13 08 27 01 50	15 29 39 39 43

This table was reprinted from Fisher and Yates, *Statistical Tables for Biological, Agricultural and Medical Research*, Longman, London (previously published by Oliver and Boyd, Edinburgh), with permission of the authors and publishers

Chapter Three □ Economic Aspects of Environmental Management

3.1
INTRODUCTION

The growing concern about environmental management, depletion of resources, and pollution control has stimulated a profusion of literature on the economic aspects of these problems.[1-9] A particularly good presentation is given by Freeman et al.[10] An economic approach to environmental management can be based on an attempt to find a dollar equivalent for all impacts, and then to determine an operating condition that maximizes the difference between benefits and costs.

As is discussed in Section 3.5, maximization of net benefit, that is, the difference between benefits and costs, occurs automatically for the sale and purchase of commodities when the following conditions are satisfied:

- Producing firms and consumers seek to optimize their individual utilities.
- Production costs and sale prices are known functions of the amount of commodity exchanged.
- The cost and price curves are unaltered by the decisions of a producer or consumer.
- The action of one individual does not affect the utility of another.
- All resources and goods are privately owned.

These conditions are an idealization of our actual market situation, but in many cases the model is sufficiently valid so that we are not too far from maximization of net benefit for society. (This model does not consider the distribution of costs and benefits among sectors of society, e.g., poor versus rich, and so a societal optimization may not be too satisfactory from the standpoint of the distribution of wealth. Concern for equity between groups is usually given separate consideration, such as the use of transfer payments from rich to poor and progressive income taxes.)

However with regard to many environmental problems, particularly pollution, there is great disparity between the idealized model and the actual situation. For example, the requirement that the actions of one individual not affect the utility of another is often not satisfied. The paper mill that pollutes a stream or the factory that emits chemicals into the air adversely influences the utilities of those that use the stream and breathe the air. These effects are termed *externalities*, where the actions of one firm or individual incurs damage (or benefit) upon another without appropriate compensation. Secondly, the price of environmental resources is not generally established. What, for example, is the value of not polluting a stream? Finally, environmental resources, as air and water, are not usually under private ownership, so that it is difficult to establish and collect an appropriate price for these resources.

Environmental deterioration is due, at least in part, to these *market failures*,[11] that is, to discrepancies between the ideal market situation and the actual prevailing conditions. The fact that waste production is often an externality means that the cost of damages due to waste is not a cost to the producer. There is no incentive for the paper mill, seeking to optimize its profit, to clean up a stream if it is not charged for its effluence. A lack of private ownership of resources may result in a rapid depletion of these resources. Underground oil sources are pumped intensively and fisheries are exhausted because any single individual wishes to maximize his share of the supply. *User costs*, which are the *opportunity costs* of future returns, are not given adequate consideration. (An *opportunity cost* is the loss that results from not following the next most desirable alternative course of action. Opportunity costs of future returns are losses that accrue from not having the resource available at some future date.) In response to the problems of externalities and resource depletion, it has been suggested[12-14] that government intervention is required in the form of regulations, subsidies, and taxes to provide incentives to reduce pollution. Possible formats for this intervention are considered in Sections 3.6 through 3.8.

Economic studies of environmental systems or pollution control generally begin with a consideration of the costs and benefits associated with various alternatives. Methods for assigning costs and benefits are discussed in Section 3.2. Evaluation of alternatives may be performed by optimizing an objective function that is the difference between benefits and costs.

3.2
COSTS AND BENEFITS

The economic approach to environmental analysis and management consists of cost determination, benefit determination, and optimization of net benefit.

Cost Determination

Labor, capital, land, and materials are required for environmental projects or pollution control. The utilization of these resources in this manner is a cost to society, presuming, of course, that there is some alternative desirable use to which these resources could be applied. Costs may involve the expenditure of funds for buildings or capital equipment, and these are termed *capital costs*. Such costs depend on the capacity of a dam, the power output of a generator, the capacity of a sewage plant, and so on. *Operating costs* are the costs of conducting business. These include wages paid to laborers, rental for land, expense of materials used in operation, and so on. Capital and operating costs are not necessarily independent. For example, a larger investment in capital equipment might result in a more efficient production process and thereby reduce operating costs. Whenever possible, cost estimates are based on the costs of previous projects of a similar nature. A difficulty of cost estimation is the unknown cost changes resulting from new technologies or procedures.

Tables 3.1 and 3.2 list the estimated costs of water and air pollution for the period 1970–1975 to achieve established air and water quality standards.[15]

Table 3.1
Costs of Water Pollution Control for the 5-Year Period, 1970–1975
(Billions of Dollars)

	Capital Costs	Operating Costs	Total
Public			
Federal	.3	1.3	1.6
State and local	13.6	9.3	22.9
Subtotal	13.9	10.6	24.5
Private			
Manufacturing	4.8	7.2	12.0
Other	.5	1.0	1.5
Total	19.2	18.8	38.0

Table 3.2
Costs of Air Pollution Control for the 5-Year Period, 1970–1975
(Billions of Dollars)

	Capital Costs	Operating Costs	Total
Federal	.4	1.2	1.6
Private	13.4	8.7	22.1
Total	13.8	9.9	23.7

In addition to the resource costs there may be costs associated with a reduced value for the factors used in production, namely, land, materials, labor, and capital. For example, if a factory closes because of the high costs of pollution control and there are no other employment opportunities, then a segment of the labor force remains unemployed. Or if someone owns land containing high sulfur-content fuel that cannot be burned because of pollution constraints, this land will produce less income. These factor costs are generally more difficult to calculate than the resource costs; in most cases they are appreciably less than resource costs, and generally they are not included in cost-benefit calculations. However, although the cumulative cost to society for factor devaluation may be less than the cumulative resource costs, the cost to the unemployed individual is very high.

Benefit Determination

The monetary benefits of an environmental project may be either a direct or indirect result of the program. A direct benefit of a dam might be measured in terms of the reduced cost over existing methods for providing water or hydroelectric power. An indirect benefit may be the increase in agricultural output that results from an augmented water supply.

There is not a unique method for measuring nonmonetary benefits, such as the maintenance of esthetic features, but one approach is to establish an indifference relationship or dollar equivalence between the benefit and a *willingness to pay*. That is, what would the consumer be willing to pay to have certain environmental amenities, even though no actual payment would occur?

Willingness to pay may be estimated in several ways. If a state government is planning a shoreline recreational facility, the willingness to pay for this would be similar to the fee for a comparable existing private facility. Or a questionnaire might be used to determine the payments individuals would be willing to make.

For example, recreational benefits associated with water-pollution control for the Delaware River were required for a cost-benefit study.[16] It was determined that existing pollution levels precluded recreational uses. Population and economic projections were performed to estimate the number of "man days" of boating and fishing that would occur if pollution control were exercised. Each man day was assigned a willingness to pay a value of about $3, whereupon a dollar value for recreational benefits could be calculated.

Table 3.3 provides an example of a cost-benefit study on a public project intended to control beach erosion in St. Johns County, Florida.[17] Much of the shore of St. Johns County had been eroded by wave action and currents, and property along the shoreline had sustained severe damage. The Army Corps of Engineers decided that the placement of fill along the coast would protect the property and also provide recreational beaches. The analysis was as follows:

Table 3.3
Beach Erosion Control Study

Estimated Initial Costs

Placement of Beach Fill	$2,446,000
Engineering and Design	120,000
Supervision and Administration	150,000
	$2,716,000
Lands, Easements, Rights-of-Way, and Relocations	
Total Initial Cost	110,000
	$2,826,000

Estimated Annual Costs

Periodic Beach Nourishment (Fill Replacement)	$ 530,000

Estimated Annual Benefits

		Non-Federal		
	Federal	Public	Private	Total
Benefits from Prevention of Damage	0	$ 35,000	$248,000	$283,000
Recreational benefits (estimate of willingness to pay)		420,000		420,000
Benefits to a Federal Navigation Project by preventing the degradation of St. Augustine Harbor	$83,000			$ 83,000
Total	$83,000	$455,000	$248,000	$786,000

The benefit-to-cost ratio asymptotically approaches a value of 1.5, and is close to this asymptote for time periods in excess of 10 years. In this example, the bulk of the benefits results from a willingness-to-pay estimate for recreational uses. A more conservative assignment of the benefits associated with recreation would significantly reduce the benefit-to-cost ratio.

A determination of willingness-to-pay is equivalent to a determination of utility, where utility is measured in dollars. As discussed in Chapter 4, a typical curve for utility as a function of the quantity Q of a commodity (quantity may refer to hours of recreation, number of automobiles, etc.) tends to have a decreasing derivative for increasing Q. Figure 3.1 illustrates this utility curve. The derivative of this curve (dU/dQ) is the utility increment per unit of the commodity, and therefore is the dollar value society assigns to an additional commodity unit. This dollar-per-unit is the *price* that society is willing to pay for another unit.

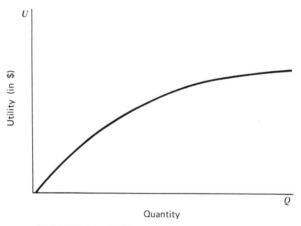

Quantity

FIGURE 3.1.   Utility as a function of quantity.

Figure 3.2 shows price (dU/dQ) obtained from the derivative of the curve of Fig. 3.1. Price as a function of Q is termed a *demand* curve, that is, it illustrates the demand for a commodity, as reflected by its price, as the available quantity varies.

If an amount Q_s is provided, as illustrated in Fig. 3.3, the revenue obtained is $P_s Q_s$. However most consumers are willing to pay more than Q_s so that the perceived benefit exceeds the revenue. From our previous

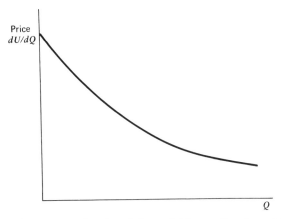

FIGURE 3.2.   Price as a function of the available quantity of a commodity.

discussion we know that price $P(Q)$ is given by

$$P(Q) = \frac{dU}{dQ},$$ (3.1)

so that the total utility for a quantity Q_s is

$$U = \int_0^{Q_s} P(Q)\,dQ.$$ (3.2)

The area contained in the $P_s Q_s$ rectangle in Fig. 3.3 is the revenue, whereas the area under the curve between 0 and Q_s is the willingness-to-pay benefit. Even if there is no revenue, that is, the selling price is zero, a willingness-to-pay benefit may still be calculated from (3.2) with $Q_s \to \infty$. In this manner, the utility derived from free recreation, maintenance of a scenic area, or protection of flora might be evaluated.

In some instances it is difficult to determine willingness to pay. For example, most individuals would be uncomfortable assigning a payment for clean air because of a lack of knowledge of the benefits associated with a specified improvement in air quality. In this case it is preferable to determine the benefits of pollution control by estimating the amount of damage avoided by a cleanup program.

Air pollution affects health, vegetation, materials, and property values. As always, there is the problem of relating incommensurate variables such as health and dollars. Relationships have been found between air quality

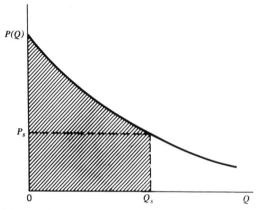

FIGURE 3.3.   Price as a function of Q, showing revenue and willingness-to-pay benefit.

and mortality rates,[18] and from these data it is possible to associate a change in mortality rate with a specified improvement in air quality. One assumption that has been made was that the dollar-value damage associated with premature death due to pollution is the amount lost in earnings. The damage associated with illness was assumed to be the loss in earnings plus the cost of treatment. On this basis, a 50% reduction in particulates and sulfur oxides yielded a $2 billion annual benefit nation-wide (as of 1963) for the dollar equivalent of health.[18]

The costs and benefits may be summarized as follows:

1. Costs
    a. Resource costs: costs of labor, materials, capital, land.
        i. Capital costs: investment in capital equipment.
        ii. Operating costs
    b. Factor costs: reduced value for factors of production.
2. Benefits
    a. Monetary
        i. Direct: such as the reduced cost of electricity from a dam.
        ii. Indirect: such as the increased agricultural output from the aug-mented water supply of a dam.
    b. Nonmonetary
        i. Willingness to pay: this is determined from the integral of the demand curve.
        ii. Avoidance of damage.

Optimization of Net Benefit

With costs and benefits all measured in dollar-equivalent units, it is then possible to define a utility or objective function that is the net benefit, i.e., the difference between benefits and costs. The independent variables in a pollution problem might be the amount of resources committed (land, labor, capital, and materials) to pollution control, the type of technology employed, and the percentage of materials recycled. Optimization of net benefit with respect to the independent variables would then give the desired operating conditions.

This is not the only possible approach. A standard may be established for pollutant level that is not to be exceeded, and then the costs of achieving this level are minimized. This obviates the requirement for determining a dollar equivalent for benefits. The standard might be fixed by health and/or technical feasibility considerations. A further discussion of different options for optimization is given in Section 1.5 with regard to methods used in cost-effectiveness analysis.

Who Pays and Who Benefits?

It has been mentioned previously that optimization of net benefit is generally performed for society as a whole, and does not distinguish between which sectors of society pay the costs and which reap the benefits. This important subject has been considered by Freeman et al.[10] Improvement in air quality is particularly important to the urban centers, which house a preponderance of lower-income families. The wealthy can move to the suburbs, where the water and air are generally purer. Water-quality improvement results primarily in additional recreational benefits, which is of more concern to the higher-income groups.

The resource costs for pollution control or a public-service project incurred by a public agency are paid by taxes or a charge for services. If taxes are collected by the federal government, this can be done progressively, that is, with higher-income groups paying a higher percentage of the tax. If a tax is collected locally by a sales tax, for example, it will be regressive, where lower-income groups pay a higher percentage. A charge for services, such as for the electricity and water output of a dam, is usually passed on to the consumer in the form of a higher price for the product that requires the service. This is also a regressive charge, for an equal increment in cost to rich and poor means a higher percentage for the lower-income category. If the resource cost is borne by private industry, this will again result in a higher price for the product.

Additional costs result from a devaluation of the factors of production, such as labor, capital, and land. Workers may become unemployed if pollution-control requirements force a company out of business. If other employment opportunities do not exist, this would be a regressive cost, for the primary burden would fall on a group with little or no income.

Freeman et al.[10] consider two approaches to alleviating regressive cost burdens: (a) a cost subsidy or (b) adjustment assistance. The cost subsidy would provide federal payments for resource costs, so that the burden is shifted from the consumer to the taxpayer. Therefore a regressive product-cost increase can become converted to a progressive income-tax assessment. However, cost subsidies do not provide incentives for the polluter to diminish his effluence by changing his product, improving his processing, recycling, or other measures. The polluter still does not identify the residuals he produces as part of his costs.

Rather Freeman et al.[10] suggest adjustment assistance, to provide aid directly to those who are adversely affected. If a worker becomes unemployed he could receive retraining and unemployment compensation; if a factory is closed, low-interest loans could be provided for reconversion, and technical assistance would be available to industries for new product development. In this manner, a more equitable distribution of the costs of pollution control could be achieved, while still maintaining incentives for pollution reduction by the polluter.

3.3
COSTS AND BENEFITS AS FUNCTIONS OF TIME

Costs and benefits of a public project or pollution control depend on both time and the magnitude of the effort. Let us consider a long-term public project, such as a dam. In the initial stage of the program, costs accrue from an investment in planning and design, whereas no benefits are derived. This time period is shown as Stage I in Fig. 3.4. The next stage involves the building of the structure and the installation of equipment. Annual costs reach a peak during this period and start to diminish as the construction is completed. The primary expenditure during Stage II is for capital costs. Benefits commence with Stage III, when the project is in operation. As the project ages, the benefits may drop because new technol-

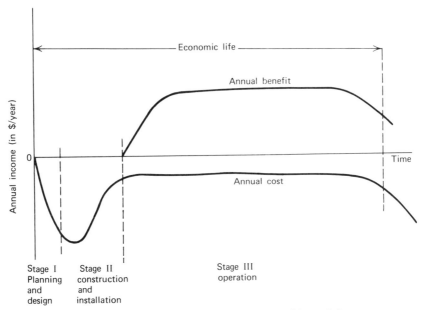

FIGURE 3.4.   Annual benefits and costs as functions of time.

ogies or other time-dependent factors (e.g., shifting populations) might reduce income. During Stage III the costs are primarily operating costs, which remain relatively constant for a long period of time, until obsolescence and aging cause an increase in annual costs. The economic life of the project is the period from initiation to the time that net annual benefits drop to zero. After this time the project loses money each year and is no longer profitable. A large project, such as a dam, may require 30 years to complete the first two stages and the operation period will be from 50 to 150 years long. Cumulative benefits and costs are obtained from Fig. 3.4 by integration, and are shown in Fig. 3.5. The economic life is the time required for the cumulative net benefits to reach a maximum.

Costs and benefits that occur at some time in the future are *discounted*, so that deferred costs and benefits may be related to their present worth to reflect available interest rates. For example, if money may be invested at an annual interest rate of 6%, the $100 received after 12 months has a present worth of $(1/1.06) \times \$100 \cong \$94$, since an immediate investment of $94 will produce $100 after one year at 6% annual interest. If p_i is the net benefit or profit derived in the ith year hence, and the annual interest rate

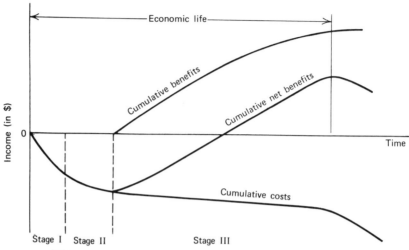

FIGURE 3.5.   Cumulative benefits and costs as functions of time.

is j, then the present worth pw of a project is

$$pw = \sum_{i=0}^{N} \frac{p_i}{(1+j)^i} \tag{3.3}$$

where N = economic life of the project in years. For $pw = 0$, the present worth of benefits is exactly equal to the present worth of costs.

From Eq. (3.3) it is apparent that the present worth may be increased or decreased by varying j. Particularly during times when the long-term interest rate is changing, a range of interest rates may be considered. Those who are inclined to favor public projects tend to suggest a low value for j, and those opposed to these projects tend to recommend a higher value.

If $p_i < 0$ in the early stages of a project, as is generally true because planning and construction costs precede benefits, and the cumulative net benefit is positive over the economic life, then it is always possible to choose a value for j sufficiently large so that $pw = 0$. The value for j that causes pw to equal zero is termed the *internal rate of return* for the project. If alternative investment interest rates exceed the internal rate of return the pw is negative and the project loses money in comparison to the investment, whereas if interest rates are below the internal rate of return the project shows a profit. Therefore, the internal rate of return is the interest rate at which the project would be abandoned on the basis of present worth being positive.

For the case that there is an initial investment loss L and an annual net benefit B extending indefinitely from the first year, then it may be shown that the internal rate of return is (B/L). From (3.3),

$$pw = -L + B \sum_{i=1}^{\infty} \frac{1}{(1+j)^i} = -L + \frac{B}{j}.$$

For $pw = 0$, $j = (B/L)$.

3.4
COSTS AND BENEFITS AS FUNCTIONS OF SCALE

As the scale of the project grows, such as the capacity of a dam or the size of a sewage-treatment plant, costs start to rise monotonically but with a decreasing derivative. A cost curve as a function of scale might be of the form shown in Fig. 3.6. The slope of the cost curve is the cost per unit increase in scale, and this derivative decreases because of the savings that result from large-scale production. The 100th car produced annually by a

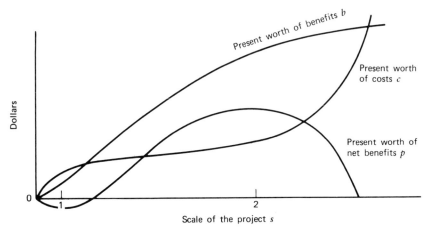

FIGURE 3.6. Costs and benefits as functions of the scale of the project. (The units for project scale might in terms of the capacity of a dam, or the quantity of waste treated per unit time by a sewage plant.)

small factory is more expensive to fabricate than the 100,000th car of comparable construction produced by a large factory. This effect of reduced cost per additional item as the scale increases is termed *economies of scale*. As the scale of the project becomes still larger, the slope usually increases, that is, the cost per additional item rises, because the operation may become more difficult to manage or special equipment may be required.

Benefits rise with the scale of the project, with a slope that decreases for large scales as the requirement for additional output diminishes in response to a saturation of market needs. The benefits and costs illustrated in Fig. 3.6 may be derived over a long time span, and it is presumed that all future returns have been referred to their present worth by discounting. Net benefits, the difference between benefits and costs, is also shown in this figure.

The economic criterion for selecting the scale of the project is the point at which net benefits p are maximized. For maximization,

$$\frac{dp}{ds} = \frac{db}{ds} - \frac{dc}{ds} = 0, \tag{3.4}$$

and points 1 and 2 in Fig. 3.6 correspond to the scale values for which Eq. (3.4) holds. The derivative db/ds is termed the *marginal benefit* with respect to scale change, and dc/ds is the *marginal cost* with respect to scale change. From (3.4) we see that at the point of maximum net benefit the marginal benefit equals the marginal cost.

If we differentiate the benefit/cost curves in Fig. 3.6 we obtain the marginal benefit/cost curves in Fig. 3.7. From this figure we note the reduction in marginal costs that results from economies of scale, until the scale becomes so large that increasing size is no longer beneficial. The point of maximum net benefit may be obtained from Fig. 3.7 by subtracting the two curves or, alternatively, from the ratio of the two curves. From (3.4), we find that at the maximum net benefit point

$$\frac{db/ds}{dc/ds} = \frac{db}{dc} = 1, \tag{3.5}$$

so that where the two curves in Fig. 3.7 have the same value we have the condition $(dp/ds) = 0$.

Figure 3.8 shows (db/dc), which is the ratio of the two curves in Fig. 3.7. At points 1 and 2, $(db/dc) = 1$, but the positive slope at point 1 indicates

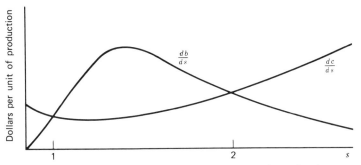

FIGURE 3.7.   Marginal costs and benefits as functions of scale.

that this corresponds to a minimum of p, whereas the negative slope at point 2 indicates a maximum for p. Therefore point 2 is the scale of the project that maximizes the present worth of net benefit. Also included in this figure is the ratio (b/c), which is often used as a figure of merit for an undertaking. In general, however, the maximization of (b/c) does not correspond to net benefit maximization, and usually occurs at a scale smaller than that for maximum net benefit.

If net benefit is a function of several variables, x, y, and z, then without constraints on the system, namely, x, y, and z are not related to each other, the maximum point is given by

$$\frac{\partial b}{\partial x} = \frac{\partial c}{\partial x} \qquad \frac{\partial b}{\partial y} = \frac{\partial c}{\partial y} \qquad \frac{\partial b}{\partial z} = \frac{\partial c}{\partial z}. \tag{3.6}$$

At the extremum, that is, the point at which net benefit is maximized, the

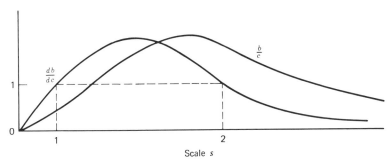

FIGURE 3.8.   Marginal benefit to marginal cost ratio and benefit to cost ratio as functions of scale. Point 2 corresponds to the scale for maximum net benefit.

marginal benefit equals the marginal cost with respect to all of the variables. The variables might include the number of laborers employed, the capital invested, and the scale of the project. At the point of maximum net benefit, for example, the marginal benefit with respect to an increase in the labor force equals the corresponding marginal cost.

3.5
SUPPLY AND DEMAND

In the previous section we observed that net benefit is maximized when marginal benefit equals marginal cost. Under certain assumptions, this condition is satisfied with regard to the production and purchase of commodities. These assumptions do not hold when residuals, as pollutants, are involved, thereby contributing to problems of environmental deterioration. It is useful to have some understanding of market exchanges and market failures with regard to pollution.

The demand curve, as illustrated in Fig. 3.2, is the price that the consuming public is willing to pay for an additional unit of commodity as a function of the available quantity. Price equals the marginal utility (or benefit) with respect to quantity.

The curve of Fig. 3.2 specifies a functional dependence between the two variables, price and quantity, and another relationship is required to determine uniquely the equilibrium values for these variables. This other relationship is termed the *supply* curve. A supply curve relates the price for which a commodity is sold to the quantity that the producers are willing to provide. In general, the higher the price, the more will be provided, so that the curve has a positive derivative.

Figure 3.9 illustrates supply and demand curves. If the quantity produced and sold is Q_1, below the intersection of the two curves, the price the consumer is willing to pay, P_1, corresponds to a higher production on the supply curve and thus the producers are inclined to increase output. If the quantity produced and sold is Q_2, above the intersection of the curves, the price the consumer is willing to pay, P_2, corresponds to a lower production on the supply curve causing the producers to reduce production. In equilibrium, the intersection determines the quantity and price.

In a freely competitive market, equilibrium for Fig. 3.9 corresponds to the condition of maximum net benefit, where marginal benefit equals marginal cost. If marginal benefit (i.e., willingness to pay for an additional

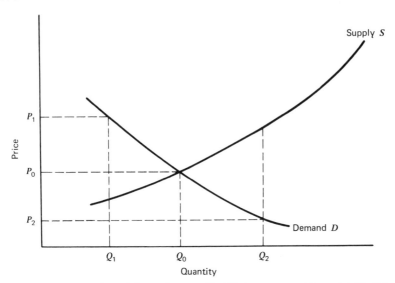

FIGURE 3.9.   Supply and demand curves. Equilibrium occurs at the point (P_0, Q_0).

unit) exceeds marginal cost (i.e., the cost to produce an additional unit), the production will increase, since this enables the producer to increase his profit. If marginal benefit is less than marginal cost, production will diminish since the producer is losing money.

For the market to operate in this manner, wherein net benefit is optimized, it is necessary that certain conditions hold. These conditions were stated in Section 3.1 and are repeated below:

• Producers and consumers seek to optimize their individual utilities.
• Prices and production costs are known functions of the amount exchanged.
• Supply and demand curves are unaltered by the decision of a consumer or producer.
• The action of one individual does not affect the utility of another.
• Resources and goods are privately owned.

When environmental resources, such as air and water are involved in the production process and externalities are generated, these conditions are not satisfied. In this case the resources are not privately owned and the action of one individual affects the utility of another. Alternatively, it may be said that the production costs are undervalued because no cost has been associated with the use of environmental resources and no cost has been assigned to the resources required to repair or avoid the damages resulting from pollution.

If these costs could be determined and added to the producer's cost, this would shift the supply curve upward, as shown in Fig. 3.10. In the absence of environmental costs, the equilibrium point is at (P_0, Q_0). If P_E, the price increment resulting from environmental costs, is charged to the producer, the new supply curve is S'. This results in a new equilibrium point (P_0', Q_0'), with a reduced quantity and a higher sales price. The increase in the sales price, $P_0' - P_0$, is less than P_E, so that a portion of P_E is borne by the consumer and a portion by the producer.

Without P_E the marginal production costs have been underestimated because environmental costs have not been included. Therefore, at the original equilibrium point (P_0, Q_0) marginal costs exceed marginal benefits. With P_E the producer will include environmental costs in his marginal cost determination and thereby reduce production until the new price, P_0'

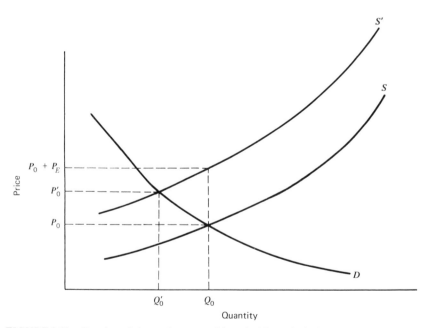

FIGURE 3.10. Supply and demand curves, with and without the inclusion of environmental costs.

equals his new marginal cost. Hence (P'_0, Q'_0) corresponds to maximization of net benefit with the inclusion of environmental costs.

We see that net benefit may be optimized when environmental costs are involved, by adding these costs to the costs of the producer. The higher the environmental costs, the greater is the reduction in the quantity sold. This means that items that produce small amounts of pollution would be substituted for the high pollution items. In addition, producers would have an incentive to introduce new technology or substitute products that result in fewer undesirable residuals so as to reduce their cost.

It has been suggested that this market failure, namely, the lack of inclusion of environmental costs, may be corrected by having the government impose an *effluence tax* on the producer.[12] This tax would charge the producer in proportion to the mass of effluence emitted. Ideally, the tax would be equal to the cost associated with the generation of residuals, so that net benefit would be optimized. In addition to the effluence tax, a number of other techniques for pollution control have been proposed, and it is useful to consider some of the advantages and disadvantages of alternative courses of action.

3.6
METHODS FOR ENVIRONMENTAL PROTECTION

In the previous section it was noted that an effluence charge could be used to maximize net benefit. However it would be difficult to determine the exact amount of this charge, since the costs of pollution control are not known precisely. Other possible approaches are:

1. An appeal to the civic responsibility of the polluter to reduce emissions. This appeal could be strengthened by pressures from community organizations that are concerned about the problem. This approach does not require any legislation nor any administrative structure. However pollution reduction costs money, and without economic incentives or pollution taxes it is improbable that the producer will respond to civic concern.

2. Adversary proceedings might be initiated wherein an individual or group that has suffered damage from pollution can seek compensation from the polluter. In many instances, however, it would be difficult to establish both the origin and amount of the damage. For example, an

individual residing in an area with excessive air pollution may suffer health damages, but it is difficult to determine the dollar equivalent of this damage and to identify a specific source.

3. Direct regulation may be imposed, such as the limitation of pollutant emissions to certain specified levels. These regulations could be enforced by a licensing procedure that would require an acceptable emission level as a condition for licensing. The permitted effluence emission level would be determined by relating emission to the desired standards for environmental quality.

However establishing a standard identical for each polluter is not usually the least expensive approach to achieving an overall level of quality. This means that the cost to the consumer will be higher than necessary. The total cost C for pollution removal is the sum of the costs C_i, for each polluter:

$$C = \sum_i C_i(Q_i), \tag{3.7}$$

where Q_i is the quantity removed by the ith polluter. To achieve a specified level of environmental quality, this means that the total amount removed Q is given, where

$$Q = \sum_i Q_i. \tag{3.8}$$

If cost [Eq. (3.7)] is to be minimized with the constraint, [Eq. (3.8)], this is equivalent to (see Appendix B):

$$\text{Minimize}_{Q_i} \left\{ \sum_i C_i(Q_i) - \lambda \left[\sum_i Q_i - Q \right] \right\}, \tag{3.9}$$

The minimization of (3.9) gives

$$\frac{\partial C_i}{\partial Q_i} = \lambda = \frac{\partial C_j}{\partial Q_j}. \tag{3.10}$$

This means that if cost is to be minimized, the marginal cost with respect to the amount removed must be the same for each polluter. If there is a uniform standard, so that each polluter must remove a fixed amount or fixed percentage of his effluence, it would be an unusual coincidence if this corresponded to equal marginal costs. It will be shown in the following section, however, that an effluence tax does lead to equal marginal costs for all polluters. In Section 3.10 a numerical example is presented which illustrates the cost difference between a uniform standard and an effluence charge.

3.7
THE EFFLUENCE TAX

In a laissez-faire system, a polluter has no economic incentive to reduce his effluence, for to do so would increase the cost of his product without any comparable benefits. But benefits do result from pollution reduction, in terms of the damage that is avoided. If these benefits are transferred, in some manner, to the polluter then he has an incentive to reduce his effluence. An effluence tax performs this function.

The benefits to society that are derivable from pollution control, namely, the damage avoided, as a function of the percentage of pollutant removed, might be as shown in Fig. 3.11. This curve increases monotonically but need not be linear, since the damage reduction is not necessarily linearly proportional to the percentage of waste removed.

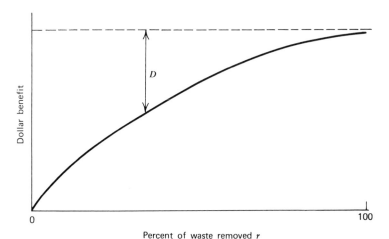

FIGURE 3.11.   Benefits of waste removal as a function of the percentage of waste removed. D is the damage that results from the emission of pollutants.

The benefit of pollution control may be assigned to the polluters by means of an effluence charge equal to the damage D shown in Fig. 3.11. By reducing his effluence, a polluter would then derive a benefit equal to the societal benefit. An effluence tax, that is, a charge per unit of effluence emitted, should equal the ratio of monetary damage to the quantity emitted. If the curve in Fig. 3.11 is not a straight line then this ratio is not constant, but it is simpler to assign a fixed effluence tax T_E, as shown in Fig. 3.12.

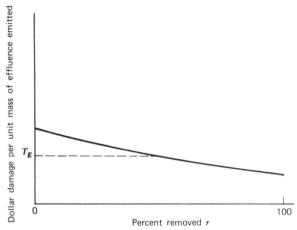

FIGURE 3.12. Dollar damage per unit mass of effluence emitted as a function of the percentage of waste removed.

How will the polluter react to an effluence tax? If the tax exceeds his cost per unit of waste removal, he will reduce his effluence to the point where his marginal cost for waste removal becomes equal to the tax. A typical cost curve for pollution reduction for a single polluter is shown in Fig. 3.13. This cost curve is similar to that in Fig. 3.6, and for the same reasons. The derivative of the cost curve reduces in response to economies of scale, and then increases because high levels of purification require expensive procedures. With a fixed effluence tax T_E the total cost C to the polluter for removal of r percent of his effluence is

$$C = \left(1 - \frac{r}{100}\right) Q_T T_E + C_R(r), \tag{3.11}$$

where $C_R(r)$ is the cost of removal, and Q_T is the total quantity of

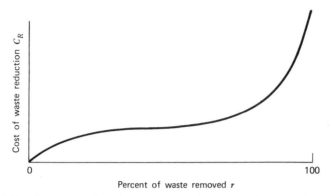

FIGURE 3.13.   Cost to a single polluter for waste reduction, as a function of the percentage of waste removed.

pollutant generated, so that

$$\frac{r}{100} Q_T = Q, \tag{3.12}$$

is the quantity, Q, that is removed. To minimize cost, the condition is

$$\frac{dC}{dr} = 0 = -\frac{Q_T T_E}{100} + \frac{dC_R}{dr}. \tag{3.13}$$

Or substituting (3.12) into (3.13),

$$\frac{dC_R}{dQ} = T_E. \tag{3.14}$$

Equation (3.14) indicates that at the minimum cost condition, the marginal cost with respect to the amount reduced should equal the tax. This point is illustrated by the intersection of the curves in Fig. 3.14. The percent removed is r_0. If there is no minimum in the range $0 < r < 100$, then the minimum cost occurs either at $r = 0$ or $r = 100$. For example, if the effluence tax is T'_E, as shown in Fig. 3.14, then the minimum cost is at $r = 0$ since the tax is always less than the marginal cost. If there are two intersections, such as for T''_E, it is necessary to evaluate (d^2C/dr^2). A minimum occurs for $(d^2C/dr^2) > 0$ at r''_2, whereas there is a maximum for $(d^2C/dr^2) < 0$ at r''_1.

Equation (3.10) indicates that when many polluters are involved, the minimum total cost for a specified amount of pollution removal occurs

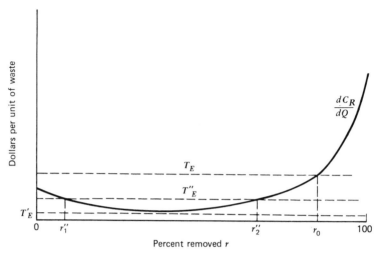

FIGURE 3.14. Marginal costs as a function of the percentage removed, with various effluence taxes, T_e, T'_E, and T''_E.

when all the marginal costs for each of the individuals are equal. An effluence tax achieves this condition. Since all polluters are confronted with the same effluence tax, from Eq. (3.14) we have that the marginal costs are all equal. Therefore, an effluence tax results in the lowest cost for obtaining a specified amount of waste reduction.

Figure 3.15 shows the amount paid by the polluter. If the polluter removes a percentage specified by r_1, as indicated in Fig. 3.15a, it costs him an amount proportional to A_1 for waste reduction and an amount proportional to A_2 must be paid to the government as an effluence tax. The total cost C is

$$C = \frac{Q_T}{100}[A_1 + A_2]. \qquad (3.15)$$

The polluter is not concerned about how much of his cost results from A_1 or A_2, but rather he wishes to minimize the sum $(A_1 + A_2)$. This minimum occurs at r_0, as shown in Fig. 3.15b.

Constant utility lines provide an alternate method for finding r_0. The solid lines in Fig. 3.16 are lines of constant utility, given by

$$A_1 + A_2 = \text{constant}.$$

The dashed line in this figure is a relationship between A_1 and A_2 obtained from Fig. 3.15. That is, for every value of A_1 there is a corresponding value

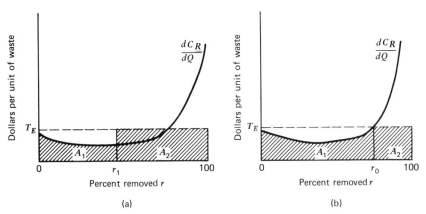

FIGURE 3.15.   The cost for waste reduction proportional to A_1 and the cost for an effluence tax proportional to A_2 for two different values of r.

of A_2, and this functional dependence is shown as the dashed line in Fig. 3.16. At the tangency point T, maximum utility occurs, specifying A_1 and A_2 for maximum utility. The percentage removed r_0 is obtained from Fig. 3.16 by finding the value of r corresponding to A_1 and A_2 at point T.

The polluter has two components to his cost: (a) $Q_T A_1/100$, which is the cost of waste treatment, and (b) $Q_T A_2/100$, which is the effluence tax

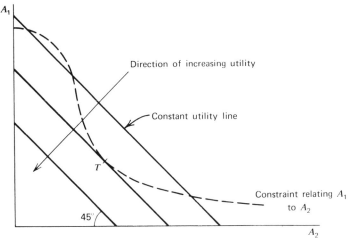

FIGURE 3.16.   Constant utility lines for waste removal. Minimum cost occurs at the tangency point T. A_1 is proportional to the cost of waste reduction, and A_2 is proportional to the amount paid as an effluence tax.

payment. However only the former is a cost to society. The latter is a transfer payment from the polluter to the government, and these funds may still be used for any productive purpose. The waste treatment cost is for the resources used in pollution reduction, and these resources (labor, material, and capital) cannot be used elsewhere. Therefore in computing the cost to society of waste treatment, only $Q_T A_1/100$ is included.

3.8
OPTIONS FOR POLLUTION CONTROL

If incentives are provided for pollution reduction, a number of options are available to accomplish this. These are:

1. To reduce the amount of materials used by the production sector of the economy.
2. To treat *residuals*, where *residuals* are the byproducts of the production process.
3. To locate effluence discharge so as to minimize damage.
4. To augment the assimilative capacity of the environment.

Let us consider each of these options in turn.

1. *Reduce Material Input to Production Sector*

In Section 2.4, dealing with Materials Balance, it was shown that in the steady-state the mass of effluence returned to the environment equals the mass of materials used by the production sector.[19] Therefore, one approach to pollution reduction is to reduce the mass absorbed by production. This may be accomplished in a variety of ways:

• Reduce consumption which, in turn, reduces the mass required by the producing sector. This means a reduction in the Gross National Product (GNP), the value of goods and services for a single year.
• Recycling. The more recycling, the less need for new materials. Recycling occurs when it is less costly for the producer to recycle than to purchase new materials. Thus to increase the amount of recycling improved technologies should be studied for lowering the cost, and a higher price should be assigned to raw materials.
• Improve processing procedures so that less effluence is generated and more efficient use is made of the input material.

- Change the composition of the GNP, to substitute low residual products for high residual products.

2. *Treatment of Residuals*

Residual mass is not reduced by treatment, for, as seen from the balance of materials analysis in Chapter 2, residual mass is determined by the input to the production sector. However one type of residual can be replaced by another. For example, the gasses from a smokestack can be trapped in a liquid or solid, reducing air pollution but necessitating another form of waste disposal. It may be simpler to dispose of one type of waste rather than another, and air pollution may be the primary problem in a given community.

3. *Minimize Damage by Appropriate Location and Time of the Discharge*

The concentration of an air pollutant at ground level can be reduced by using a high smokestack rather than a low one. The BOD concentration in a stream may be minimized by storing waste until the high-flow season for the stream.

4. *Augment Assimilative Capacity of the Environment*

The assimilation rate of BOD in a stream depends upon the oxygen concentration. Oxygen content may be augmented by aeration, that is, increasing the rate of oxygen absorption. One method for accomplishing this is to spray the water into the air.

3.9
MINIMIZATION OF WASTE TREATMENT COSTS

Let us assume that pollution reduction will be accomplished by waste treatment, and that several different technologies are available. The polluter will presumably choose that combination of processes that is the least expensive to him. This will result in a higher percentage r of waste removed, and at a lower cost. For example, to maintain a specified level of cleanliness in a stream, the waste may be treated before discharge into the stream, or the effluence may be injected into deep wells that return the waste to the ocean underground, without polluting the stream. Therefore the total cost may be divided between deep-well injection costs C_1 and waste-treatment plant costs C_2. The choice of how to divide the cost between the two methods will depend on the quantity of waste Q to be reduced, where

$$f(C_1, C_2) = Q. \tag{3.16}$$

Equation (3.16) states that the quantity of waste reduced is some function $f(C_1, C_2)$ of the investments in the two methods for waste handling.

Figure 3.17 shows the lines of constant utility on the $C_1 - C_2$ plane, along with constraint lines relating C_1 to C_2 for different waste quantities Q_1, Q_2, and Q_3. For each value of Q the maximum utility is given by the point of tangency with the constant utility lines. As the waste handling capacity increases the growth line is through the points of tangency.

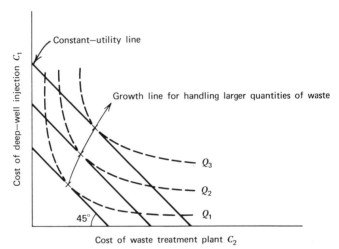

FIGURE 3.17.   The operating points for handling different quantities of waste Q_1, Q_2, and Q_3 are shown as points along the growth line.

If many alternative waste reduction methods are available, the utility optimization problem is

$$\left. \begin{array}{c} \text{Minimize: } \sum_{i=1}^{n} C_i \\ \text{With: } f(C_1, C_2 \cdots C_n) = Q \end{array} \right\}, \tag{3.17}$$

where C_i is the cost associated with the ith method and there are n

alternatives. The solution to (3.17) leads to the condition (refer to Appendix B).

$$\frac{\partial f}{\partial C_1} = \frac{\partial f}{\partial C_2} = \cdots = \frac{\partial f}{\partial C_n}. \tag{3.18}$$

Since $f(C_1 \cdots C_n) = Q$, Eq. (3.18) states that at the point of minimum cost the marginal waste reduction per investment dollar is the same for each method of waste handling.

Suppose, for example, 10^4 tons of waste are to be treated, and by deep-well injection it is possible to handle $10^{-3}C_1$ tons, and by waste treatment it is possible to handle $4C_2^{1/2}$ tons. The costs of each method are C_1 and C_2, respectively. Therefore, the problem is to minimize $(C_1 + C_2)$ with the constraint

$$f(C_1, C_2) = Q$$

or

$$10^{-3}C_1 + 4C_2^{1/2} = 10^4. \tag{3.19}$$

This is equivalent to the minimization of (see Appendix B).

$$C_1 + C_2 - \lambda \left[10^{-3}C_1 + 4C_2^{1/2} - 10^4 \right], \tag{3.20}$$

which gives

$$\begin{aligned} 1 - 10^{-3}\lambda &= 0 \\ 1 - 2C_2^{-1/2}\lambda &= 0 \end{aligned}. \tag{3.21}$$

The solution to Eq. (3.21) is

$$\lambda = 10^3, C_2 = \$4 \times 10^6,$$

and from Eq. (3.19) we have that

$$C_1 = \$2 \times 10^6.$$

To minimize waste-handling costs the investments should be in the amounts given above.

In Appendix B it is shown that

$$\lambda = \frac{\partial (C_1 + C_2)}{\partial Q},$$

evaluated at the minimum cost condition. The parameter λ, therefore, is the change in the cost per unit waste treated when the cost is minimized. Lambda has the dimensions of price, that is, cost per unit of waste, and is referred to as the *shadow price*. In this example, $\lambda = 10^3$, which means that at the minimum cost condition it requires an additional investment of $\$10^3$ for each additional ton of waste to be handled.

From Eq. (3.19) we have that

$$\frac{\partial f}{\partial C_1} = 10^{-3} \text{ and } \frac{\partial f}{\partial C_2} = 2C_2^{-1/2},$$

which, for $C_2 = \$4 \times 10^6$, gives

$$\frac{\partial f}{\partial C_1} = \frac{\partial f}{\partial C_2} .$$

This equation agrees with the requirement that the marginal waste reduction per investment dollar is the same for each method at the minimum cost condition.

3.10
COSTS OF POLLUTION CONTROL

In this section we shall compare the costs of achieving a given percentage of waste reduction by different methods of control. (This example is based on an example given in one of the references.[10]) The comparison will be made for the following alternatives: (a) an effluence tax, (b) that all polluters be required to have the same treatment percentage r, and (c) that all polluters be required to reduce their effluence by the same amount.

Suppose there are four polluters, each producing the waste mass given in Table 3.4. This table also includes a listing of the marginal costs of waste reduction for each polluter. Marginal costs are assumed constant, that is, independent of the quantity of waste treated.

If it is decided that the effluence mass discharged is not to exceed 600 kg, this may be accomplished by an effluence tax of 7 cents/kg of effluence. Then polluters 1 and 3 will eliminate their waste because the marginal cost of reduction is less than the tax. This will limit the effluence

Table 3.4
An Example of Waste Generation and Marginal Costs

Polluter	Mass of Waste Produced (kg)	Marginal Costs of Waste Reduction (cents/kg)
1	100	4
2	200	8
3	300	6
4	400	10
Total	1000	

to 600 kg, as shown in Table 3.5. The total cost to society for waste reduction is $22.00. Amounts paid as an effluence tax are transfer payments and, therefore, not included in societal cost.

Table 3.5
Waste Control by an Effluence Tax

Polluter	Mass of Waste Produced (kg)	Marginal Costs of Waste Reduction (cents/kg)	Costs for Waste Reduction	Amount of Waste Discharged (kg)	Payment on Effluence Tax
1	100	4	$ 4.00	0	0
2	200	8	0	200	$14.00
3	300	6	18.00	0	0
4	400	10	0	400	28.00
Total			$22.00	600	

If each polluter is required to eliminate 40% of his effluence, this will also result in a total discharge of 600 kg. This case is shown in Table 3.6. In this case the societal cost is $31.20.

Table 3.6
Waste Control by Uniform Reduction in Effluence

Polluter	Mass of Waste Produced (kg)	Marginal Costs of Waste Reduction (cents/kg)	Costs for Waste Reduction	Amount Discharged (kg)
1	100	4	$ 1.60	60
2	200	8	6.40	120
3	300	6	7.20	180
4	400	10	16.00	240
Total			$31.20	600

A 600 kg emission may be also achieved by requiring an effluence reduction of 100 kg from each polluter. The corresponding costs are shown in Table 3.7 with the total cost equal to $28.00. We observe that the effluence tax provides the least cost alternative.

Table 3.7
Waste Control by Uniform Treatment

Polluter	Mass of Waste Produced (kg)	Marginal Costs of Waste Reduction (cents/kg)	Costs for Waste Reduction	Amount Discharged (kg)
1	100	4	$ 4.00	0
2	200	8	8.00	100
3	300	6	6.00	200
4	400	10	10.00	300
Total			$28.00	600

3.11
SUMMARY AND CONCLUSIONS

The economic approach to environmental projects or pollution control is an attempt to specify the dollar equivalents for costs and benefits, and to maximize net benefit. Costs are associated primarily with the resources required for the project, and benefits may be measured in terms of the product output, the avoidance of damage, or a willingness to pay. Costs and benefits are both functions of the time and scale of the effort.

Under certain conditions, the sale and purchase of commodities results in the optimization of net benefit, corresponding to the point of intersection of supply–demand curves. However when residuals are involved, this optimization does not occur for free-market interactions. This market failure may be corrected by the imposition of an effluence tax. This approach is less costly than the specification of a fixed amount or percentage of pollution reduction.

The net benefit optimization achieved by the market is for society as an aggregate and does not consider the distribution of wealth. An effluence tax on private industry results in a price increase, which is a regressive imposition on the consumer. That is, a given price change is a higher percentage of the income of the poor than that of the rich. Thus if the burden of pollution control is not to be carried primarily by the poor, in addition to an effluence tax it is necessary to transfer wealth from the rich to poor. This may be accomplished by an appropriate revision of the income tax.

Several conclusions and comments regarding pollution control are:

- The least cost approach to pollution control is a residual charge.
- Improved methods are needed for the determination of environmental costs and benefits.
- Environmental management should be on a regional or geologic basis to take advantage of economies of scale. In most instances, management is on the basis of a political jurisdiction or a single manufacturer.

Some obstacles to successful environmental management are:

- The time constants associated with management are often longer than the time constants important to elected legislators. A primary concern of an elected official is to be reelected, and so there is a reluctance to impose restrictions or taxes without comparable benefits within the duration of office.
- The affluent have a disproportionate influence on the government, which inhibits the achievement of an equitable distribution of the costs of environmental management.
- The present legal system is not designed to facilitate redress for injury due to environmental degradation.

Perhaps the existence of a number of apparent crises, which now seem to confront our society, will result in the diminuation of these obstacles, and thereby clear a path to more successful management.

REFERENCES

1. T. D. Crocker and A. J. Rogers, III, *Environmental Economics*, Dryden, (1971).
2. M. I. Goldman (ed.), *Ecology and Economics: Controlling Pollution in the 70's*, Prentice Hall, Englewood Cliffs, N.J. (1972).
3. E. G. Dolan, Tanstaafl, *The Economic Strategy for Environmental Crisis*, Holt, Rinehart and Winston, New York (1974).
4. A. V. Kneese and B. T. Bower (eds.), *Environmental Quality Analysis*, Johns Hopkins Press, Baltimore (1972).
5. J. C. Headly, "The Economics of Environmental Quality," *J. Environ. Qual.* 1, (4), 377–381 (Oct.–Dec. 1972).
6. J. H. Dales, *Pollution, Property and Prices*, Toronto U.P. (1968).
7. Leontief, Wassily, "Environmental Repercussions and the Economic Structure: An Input–Output Approach," *Rev. Econ. Stat.* 52 (3), 262–271 (1970).

8. R. U. Ayres and A. V. Kneese, "Production, Consumption and Externalities," *Amer. Econ. Rev.* **59** (3), 272–297 (1969).

9. R. D. Wilson and D. W. Minnott, "A Cost/Benefit Approach to Air Pollution Control," *J. Air Pollut. Control Assoc.* **19**, 303–308 (1969).

10. A. M. Freeman, III, R. H. Haveman, and A. V. Kneese, *The Economics of Environmental Policy*, Wiley, New York (1973).

11. F. M. Bator, "The Anatomy of Market Failure," *Quart. J. Econ.* **72**, 351–378 (1958).

12. A. M. Freeman, III and R. H. Haveman, "Residuals Charges for Pollution Control: A Policy Evaluation," *Science* **177**, 322–329 (1972).

13. R. M. Solow, "The Economist's Approach to Pollution and its Control," *Science* **173**, 498–503 (1971).

14. M. O. Stern, R. Y. Ayers, and J. C. Saxton, "Tax Strategies for Industrial Pollution Abatement," IEEE Trans. on Systems, Man and Cybernetics, SMC-3, 6, 588–303 (Nov. 1973).

15. The President's Council on Environmental Quality, *Environmental Quality—1971*, p. 111, Washington, D.C. (1971).

16. P. Davidson, F. G. Adams, and J. Seneca, "The Social Value of Water Recreational Facilities Resulting from an Improvement in Water Quality: The Delaware Estuary," in A. V. Kneese and S. C. Smith (eds.), *Water Research*, Johns Hopkins Press, Baltimore, 1966), pp. 175–224.

17. U.S. Senate Document No. 97, 89th Congress, 1966.

18. L. B. Lave and E. P. Seskin, "Air Pollution and Human Health," *Science* **169**, 723–733 (Aug. 21, 1970).

19. A. V. Kneese, R. U. Ayers, and R. C. D'Arge, *Economics and the Environment, A Materials Balance Approach*, Johns Hopkins Press, Baltimore (1970).

PROBLEMS

3.1 For the numbers given in Tables 3.1 and 3.2, determine the present worth for the total costs of air and water pollution control. Capital costs occur at the initiation of the project, and operating costs are equal in each of the 5 years. Assume an interest rate of 6%.

3.2 What is the internal rate of return for the Beach Erosion Study given by Table 3.3?

3.3 A possible method for correcting the market failure with regard to residuals is to impose a consumer sales tax on commodities for which residuals are generated in the manufacturing process. Show the effect of such a sales tax on the supply and demand curves. In drawing these curves, let the ordinate be the price before the sales tax is added. Where is the new equilibrium point, i.e., what happens to the price and quantity of the commodity? Compare the sales tax approach with the effluence tax on producers with regard to the following: (a) the incentive to substitute commodities that generate fewer residuals (b) the incentive to reduce the effluence on the part of the manufacturer by waste removal or by developing new technology, (c) optimization of net benefit, that is, whether marginal costs equal marginal benefits at the new equilibrium condition, (d) which elements of society bear the burden of the costs of pollution control, (e.g., whether the costs are imposed regressively, and (e) the ease of administration and collection of the tax. Do you find significant differences between the sales tax and effluence tax approaches?

3.4 With regard to Fig. 3.15, the object of the polluter is to minimize the area $(A_1 + A_2)$. Show, using the method of Lagrange multiplier (refer to Appendix B), that minimization of $(A_1 + A_2)$ under the constraints that A_1 and A_2 are specified according to Fig. 3.15, leads to Eq. (3.14):

$$\frac{dC_R}{dQ} = T_E \tag{3.14}$$

Show that at the optimum point $(dA_1/dA_2) = -1$. Is this result consistent with Fig. 3.16?

3.5 If a firm has N employees, the cost of operation C is given by

$$C = C_0 + C_1 N$$

where C_0 and C_1 are positive constants. The revenue r as a function of N is given by

$$r = r_1(1 - e^{-r_2 N})$$

where r_1 and r_2 are positive constants. By equating marginal costs to marginal benefits, determine the number of employees that should be hired as a function of the constants in the above equations.

3.6 A questionnaire was circulated to residents in an area to determine their interest in a state park facility. Based on this questionnaire, an estimate was made of the annual number of visitors to the park as a function of the entrance fee that would be charged, as given in the following table:

Entrance Fee for an Auto ($/visit)	Number of Autos per year
8	25,000
5	100,000
2	200,000
0	300,000

(a) If it is decided not to charge a fee, determine a willingness-to-pay benefit. State your assumptions in making this calculation.

(b) If the initial cost of the project is $1.5 million and the annual maintenance cost is $280,000, determine the internal rate of return, using the willingness-to-pay benefit from (a) above.

3.7   A resource essential to daily living becomes in short supply, and several courses of action are suggested: (a) rationing with price control, (b) allowing the market price to vary without government interference, and (c) imposing a consumer sales tax. With regard to each of these courses, how do you feel that the burden of the shortage will be distributed in the society as a function of the individual's income? For example, are the costs likely to be imposed in a progressive or regressive fashion?

3.8   The damage due to a pollutant is estimated to be $40/t (per ton), and an effluence tax in this amount is imposed on a manufacturer. There are two methods of waste treatment, wherein D_1 dollars invested in method (1) results in a treatment of $.4D_1^8$ t, and D_2 dollars invested in method (2) results in a treatment of $10^2 D_2^4$ t. The manufacturer's total emissions in a specified period is 2×10^4 t.

(a) If only method (1) is used, determine the following: (i) how many tons are treated, (ii) the cost to the manufacturer, and (iii) the societal cost (include both resource and damage costs).

(b) Answer the questions in (a) above for method (2)

(c) Answer the questions in (a) if the manufacturer uses both methods in a manner to minimize his costs.

Chapter Four □ Decisions, Decisions

4.1
INTRODUCTION

In this chapter a number of different decision problems will be considered. The objective of decision analysis is to assist the decision maker to choose between alternative courses of action. Inputs to the decision process are the relationships between variables (i.e., a knowledge of the consequences of one's actions) and the priorities of the interested parties. Priorities are frequently given in general terms, and part of the decision analysis is to establish an analytic encoding for preferences.

Should the Systems Analyst Consider Normative Questions?

As noted in Chapter 1, a "most desirable" designation assigned to an alternative is a scalar description, and therefore the multidimensional attributes (e.g., cost, time, safety, and comfort) must be reduced to a scalar. This necessitates the introduction of the value systems of the interested parties, for a comparison between attributes measured in different units (e.g., cost and time) is a matter of personal preference. A 10-min reduction in time to a commuter might have a higher dollar value for one person than for another. Even when there is only one attribute (e.g., cost), the selection of an alternative may still require the inclusion of personal values, particularly when risk is involved. For example, a $100 bet to play a 50–50 gamble for $200 or zero dollars may be acceptable to one person and unacceptable to another.

Almost every decision process has an objective component, which is derived from the modeling and relates decisions to consequences, and a subjective component, which is derived from the value systems of the interested parties. There is disagreement among systems analysts (and

129

others) as to whether or not it is appropriate for the analyst to incorporate the subjective element into his study. One approach is for the analyst to present the results of the analysis without subjective consideration to the interested parties and then withdraw from the decision process. This avoids the often difficult problem of an explicit determination of personal values and allows the concerned individuals to use their intuition, judgment, and opinions to decide on a course of action. On the other hand, the analyst may seek to determine the priorities of his clients and use these priorities to assign a relative emphasis or weighting to the various attributes. In this manner, a single alternative can be recommended. An intermediate path is to present a recommended alternative for each of many different weighting schemes.

Each approach has advantages and disadvantages. Some people are "turned off" by any attempt to extract and quantify their personal preferences. "I do not want to be reduced to a number in a computer" is a common type of response. An effort by the systems analyst to incorporate the values of the interested parties may result in a negative attitude toward the entire project. In addition, it is often difficult to determine personal values, and the values that are elicited may not reflect attitudes used for decision making. That is, value information obtained from a questionnaire or interview does not necessarily indicate behavioral responses for the actual decision situation. In addition, with many groups involved in the project it may be difficult, if not impossible, to arrive at a compromise concerning a mutually acceptable set of values.

Finally, the quantification of preferences usually includes certain behavioral assumptions that may not correspond to actual behavior. For example, one assumption is *transitivity*; that is, if outcome A is preferred to outcome B and if outcome B is preferred to outcome C, then outcome A is preferred to outcome C. Symbolically, this is written as

$$\text{if}\quad A > B \text{ and } B > C$$
$$\text{then}\quad A > C.$$

Transitivity is a reasonable assumption, and one might argue that it is necessary to guarantee consistency in one's decision-making process. However people do not always exhibit transitivity. A possible explanation is that an individual does not associate a fixed value with a specified outcome, but rather changes this value depending on the available alternative outcomes and risks associated with his decisions. For example, if one is confronted with a gamble of winning d_1 dollars with probability p, or d_2 dollars with probability $1 - p$, the value associated with d_1 dollars may be functionally dependent on d_2 and p, as well as on d_1. If this is true, apparent intransitivities may occur. Whatever the reason, if the systems analyst wishes to reflect the values of his clientele he should account for this "inconsistent" behavior.

On the other hand, there are numerous reasons why it may be desirable to include subjective values in the systems analysis. In the absence of an explicit value system the decision process might be unduly influenced by the persuasiveness of a demagogue, the pressure of a lobbyist, or the weariness and desperation of the interested parties. It may be easier to detect inconsistencies in decision making when individual preferences are openly discussed and an effort is made to quantify the worth and importance of possible outcomes. It is probably more likely that those variables that are difficult to quantify will receive less emphasis without an explicit attempt to assign a weighting to the variables. Variables such as cost, time, and numbers of people are easier to grasp than factors such as esthetics and privacy, and so in the absence of a weighting scheme the latter may not receive due consideration.

The appropriate procedure to follow, whether or not to include normative values, depends on the attitudes of the interested parties, the type of problem, and the inclinations of the systems analyst. The remainder of this chapter will be devoted primarily to methods for including subjective information in the analysis.

Formats for Presenting Information

Before proceeding with this discussion it is useful to suggest several techniques for presenting information when the results of the analysis and modeling are given without the inclusion of subjective values. Table 1.6 and Fig. 1.8 are formats that may be used for describing the performance properties of the various alternatives. If one alternative is *dominated* by another (i.e., if one alternative is outperformed with regard to *all* attributes by another), it should be indicated that this alternative may be dropped regardless of the subjective positions of the interested parties. It is unlikely, however, that too many alternatives can be eliminated on the basis of dominance when a large number of attributes are involved. Some alternatives may be discarded on the grounds that one or more of their performance characteristics lie outside an acceptable performance range. This requires a subjective opinion of what is acceptable, but does not require a detailed determination of utility values.

A popular method for presenting results is to use the output from a computer model to display performance properties, such as the cost/effectiveness curves in Fig. 1.12, on a cathode-ray tube. The decision maker can then change decision-variable values and observe the resulting changes in the cost/effectiveness curve. This technique enables the decision maker to obtain rapid estimates of tradeoffs and the consequences of his actions. Also, some individuals find this approach to be more enjoyable than reading graphs, tables of numbers, or verbal descriptions.

It may be useful to indicate the best possible performance (i.e., dominant solution) for pairwise consideration of attributes. That is, the behavior of one attribute is plotted as a function of another as alternatives and decision variables are changed. Dominated performance characteristics are not included. For example, with regard to cost and effectiveness, from Fig. 1.11 we have that the dominant performance is as shown in Fig. 4.1. This curve tells the decision maker the best possible achievements for cost and effectiveness. For $C < C_0$ and $E < E_0$, alternative I dominates, and for $C > C_0$ and $E > E_0$, alternative II dominates. The choice of strategy (to impose a constraint on cost, to maximize E/C, etc.), and therefore the operating point, remains with the decision maker. Only two attributes are considered simultaneously because it would require a nonplanar figure to illustrate three or more attributes simultaneously.

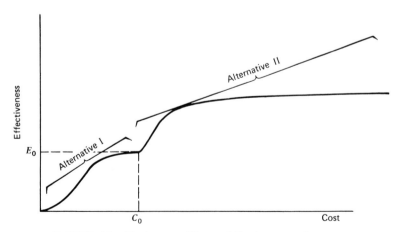

FIGURE 4.1.   The best possible cost/effectiveness performance.

Utility

Suppose it is decided to include attitudinal information in the analysis, and therefore utility ratings are sought. As discussed previously in Section 1.5, the utility rating is a number assigned to an outcome, performance characteristic, or state of the system that represents the individual's perceived worth estimate of this outcome. Utility values have been considered from both theoretical and empirical viewpoints. The former refers to philosophical considerations of attitudes, and the latter to a measurement of utility.

Over 200 years ago, Bernoulli considered the utility of wealth and decided that a given increment in wealth did not necessarily provide a fixed increment in happiness or utility. Rather, the more wealth one had, the less would be the added utility for a specified amount of wealth. A millionaire would presumably derive less pleasure from a gift of $100 than would a pauper. Bernoulli hypothesized that the utility increment is inversely proportional to one's wealth, thereby giving the relationship

$$du = \frac{b}{x} \, dx, \tag{4.1}$$

where u = utility, x = wealth, and b is a proportionality constant. Integration of (4.1) yields $u = b \ln x + c$, where c is the constant of integration. Frequently, one sees this relationship written with $b = 1$ and $c = 0$, giving

$$u = \ln x. \tag{4.2}$$

If we choose $b = \log_{10} e$, then $u = \log_{10} x$, which is an equally satisfactory function.

Contemporaries of Bernoulli agreed that utility increases less than linearly with wealth, but assumed other formulations such as

$$du = \frac{1}{x^2} \, dx \quad \text{and} \quad du = \frac{1}{\sqrt{x}} \, dx.$$

Perhaps it is reasonable to assume that an increment in utility is a function of one's utility rather than wealth. The amount of pleasure provided by a gift of $100 is dependent on one's state of happiness rather than wealth. Let us hypothesize that the utility increment is proportional to the monetary increment and to the amount of utility that is undispensed. That is, if one is completely satisfied, an increment in wealth provides no additional amount of happiness. On this basis, we are led to

$$du = b(1 - u)dx, \qquad (4.3)$$

where $u = 1$ is considered to correspond to complete satisfaction. Upon integration we obtain

$$u = 1 - e^{-bx}, \qquad (4.4)$$

where $u = 0$ has been assigned to $x = 0$. The function given by (4.4) also grows at less than a linear rate.

These hypothesized utility functions may be used to indicate attitudinal preferences or behavior patterns. Of course, it is desirable to obtain an empirical check on utility, as described in the following sections.

Chapter Outline

The remainder of the chapter is devoted to a consideration of various types of decision problems. Section 4.2 deals with a single attribute where *risk* is involved; that is, the various possible outcomes of a decision occur with known probabilities. Decision making under *uncertainty* is discussed in Section 4.3, where uncertainty means that the outcome probabilities are unknown. (A third category of decision making, decision making under *certainty*, might be considered a special case of risk, where the outcome probabilities are either unity or zero.) A number of different strategies for reaching a decision are considered in this section.

Sections 4.4 through 4.5 deal with various aspects of obtaining equivalence relationships between incommensurate variables, and several examples of equivalence relationships are presented in Section 4.4. Tradeoffs between pollution level and consumption are discussed in Section 4.4, as an illustration of utility optimization in the presence of a constraint between the variables. Finally, Section 4.5 describes a multivariable situation, where a linear weighting system is used to arrive at a utility function.

4.2
A SINGLE ATTRIBUTE WITH RISK

In this section we consider decision making when there is only a single attribute, whereas multiattribute problems are discussed in the remaining sections of this chapter. When *risk* is involved, that is, when there are a number of possible outcomes with known probabilities of occurrence, then decision analysis can play a role even for a single attribute problem.

Determination of Utility

Money will be our attribute, for it is a critical factor in many decisions. The problem is to determine courses of action when money is to be risked and several outcomes may occur. Let us assume that the possible range of costs for a project is 6×10^5 to 10×10^5. With minimum cost as an objective, the ideal is a 6×10^5 expenditure and the least desirable is a 10×10^5 expenditure. A utility u_6 associated with a 6×10^5 cost may be set equal to unity, and u_{10}, corresponding to a 10×10^5 cost, set equal to zero. (The utilities u_6 and u_{10} may be set equal to any arbitrary numbers without altering the results derived from decision making based on utility. Unity and zero are used as the extreme values merely for the sake of convenience.)

Next, we must assign utility ratings to intermediate expenditures. To accomplish this we shall employ a fundamental postulate of utility theory; when a result R_i has a probability p_i of occurring, the utility assigned to this condition is given by the mean value of utility $\langle u \rangle$. Namely, the utility under risk is

$$\langle u \rangle = \mathbf{p} \cdot \mathbf{u} = \sum_{i=1}^{n} p_i u_i$$

$$= p_1 u_1 + p_2 u_2 + \cdots + p_n u_n, \tag{4.5}$$

where u_i is the utility associated with R_i. It has been found from studies of human values, that the mean of utility is a reasonable choice to use when an outcome is expressed in terms of probabilities of results.

If an interested party is indifferent to a choice between two events this implies that

$$\langle u \rangle_1 = \langle u \rangle_2, \tag{4.6}$$

where $\langle u \rangle_1$ is the mean utility for event 1 and $\langle u \rangle_2$ is the mean for event 2. The condition of indifference may be used to establish utility values for each of the possible outcomes R_i. Suppose, for example, event 1 consists of either an expenditure of 6×10^5 with probability p, or expenditure of

10×10^5 with probability $(1-p)$. Then

$$\langle u \rangle_1 = p u_6 + (1-p) u_{10}. \tag{4.7}$$

Since $u_6 = 1$ and $u_{10} = 0$, from (4.7) we have that

$$\langle u \rangle_1 = p. \tag{4.8}$$

If we now choose event 2 to be that an expenditure of 8.5×10^5 occurs with complete certainty, we have,

$$\langle u \rangle_2 = u_{8.5}. \tag{4.9}$$

The establishment of an indifference condition between events 1 and 2 would mean that $\langle u \rangle_1 = \langle u \rangle_2$, or

$$u_{8.5} = p. \tag{4.10}$$

Therefore, if we can find a value of p that makes the interested party indifferent between events 1 and 2 we have determined that the utility for an expenditure of 8.5×10^5 is p.

The appropriate value scale for measuring money is not necessarily linearly proportional to the expenditure. For example, suppose a given course of action resulted in a 50% probability of a 10×10^5 cost and a 50% probability of a 6×10^5 cost. From the concept that two courses of action are equivalent if the mean values of their utilities are equal, if there is a linear relationship between utility and cost an equally acceptable result would be a certain expenditure of 8×10^5, where a certain expenditure means that the probability is unity that the cost is 8×10^5. However in order to avoid a 50% chance of paying 10×10^5, the interested party might be willing to accept a certain cost of 8.5×10^5. Such a decision would be characteristic of an individual that is risk averse; that is, he would be willing to pay a little extra to avoid the possibility of a more undesirable outcome. The utility $u_{8.5}$ would have the value 0.5, since

$$\langle u \rangle = 0.5 u_6 + 0.5 u_{10} = 0.5(1) + 0.5(0) = 0.5$$

$$= u_{8.5}.$$

Curve A in Fig. 4.2 illustrates the points on the utility curve corresponding to expenditures of 6×10^5, 8.5×10^5, and 10×10^5. The calculation of

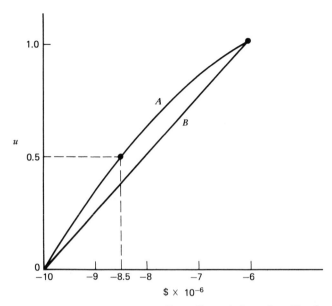

FIGURE 4.2. Utility as a function of expenditure. Curve *A* shows the utility dependence for a risk averse group and curve *B* is for a risk indifferent group. The minus sign is used along the abscissa to indicate that cost rather than income is involved.

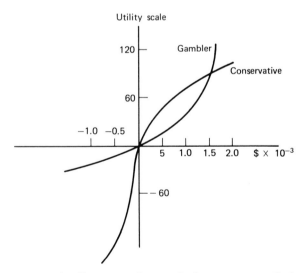

FIGURE 4.3. Measured utility curves for two business managers displaying different attitudes toward risk.

intermediate points on this curve may be determined by the same method used to evaluate $u_{8.5}$; by establishing indifference relationships for $\langle u \rangle$ between known utility values and a certainty outcome for an unknown utility value. An individual who is risk-averse would require the "odds" to be in his favor to prefer a risk situation to an outcome that is certain. Curve B is the utility relation for a group that is risk indifferent, where an increment in expenditure gives a proportional increment in utility.

Utility curves have been measured for business managers to relate their perception of the value associated with positive or negative increments of money with the actual dollar amount. Figure 4.3 shows two curves, one for a conservative manager and the other for one who is risk prone. The decisions made by these two individuals would differ significantly, as discussed in Problem 4.2.

Postulates of Utility Theory

The method that we have employed in the expenditure of funds example to obtain a value scale for a variable is based on the postulates of utility theory.[1,2] For a set of events, A, B, and C, these postulates are

(a) Transitivity.

> If $A > B$ (i.e., A is preferred to B)
> and $B > C$, then $A > C$.
> If $A \sim B$ (i.e., A is equivalent to B)
> and $B \sim C$, then $A \sim C$.

(b)   If $A > B$
> then $\left[p, A; (1-p), B \right] > \left[p', A; (1-p'), B \right]$
> only if $p > p'$,

where $[p, A; (1-p), B]$ means that outcome A has probability p and outcome B has probability $(1-p)$. This postulate means that the preference for a risky outcome increases only if the probability of a more desirable event is increased relative to the probability of a less desirable event.

(c) The act of gambling has no inherent value. This means that the compound lottery:

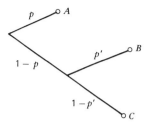

(that there is a probability p that A occurs and probability $(1-p)$ that there is another lottery with probability p' of obtaining B and probability $(1-p')$ of obtaining C)

is equivalent to the lottery:

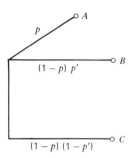

In other words, no value is associated with having an additional gamble.

(d) If $A > B > C$, there is a probability p with $0 \leqslant p \leqslant 1$ such that

$$B \sim [\,p, A\,;(1-p), C\,].$$

This means that when B ranks between A and C it is possible to establish indifference between B and a gamble involving A and C.

With these four postulates, a unique utility function may be defined for a variable. The utility function should satisfy the following conditions:

- If $A > B$, then $u(A) > u(B)$. That is, the utility for A is greater than the utility for B when A is preferred to B. (We have chosen the positive utility direction to correspond to increased satisfaction.)

- If $B \sim [\,p, A\,;(1-p), C\,]$
  then $u(B) = pu(A) + (1-p)u(C)$.

When an outcome is a combination of events with various associated probabilities, the utility for the outcome is defined as the *mean* of the utility.

In our example dealing with money, these postulates were used to develop utility scales for different possible outcomes. Once the utility function has been determined, a course of action is chosen based upon maximization of utility.

The Decision Tree[3]

A decision tree is a pictorial display of the decision process. It is a presentation of the decisions that must be made, the possible outcomes, and the probabilities associated with these outcomes. When there is a relatively small number of decisions with discrete outcomes, the decision tree is a useful device for choosing the course of action with the highest utility rating.

Suppose, for example, a manufacturer is considering the development of a new product. If the product is accepted by the consumers, he will augment his income, but if the product is unsuccessful he will have lost his development costs. The manufacturer believes that there is a 50–50 chance that his new product will sell. It has been suggested to him that he hire a firm to conduct a market survey, and from this firm's previous record he can be 90% certain that if they recommend the new product that it will be accepted. His choices, therefore, are to stay with his original line, develop the new product, or pay for a market survey and follow the recommendation suggested by the survey. Figure 4.4 illustrates this decision problem.

The net incomes listed along the right-hand side of Fig. 4.4 are assumed returns associated with each branch of the decision tree. If there is no new product development the income, in 10^4, is taken to be 1.0. If the new product is developed and successful, income is doubled, but if the product is not successful, then development costs reduce income to 0.6. It has been assumed that the cost of the market survey is $0.2 \times \$10^4$, so that the incomes with the survey are all reduced by this amount.

To choose a course of action it is necessary to specify the utilities of the outcomes and the probabilities associated with each chance outcome. Let us assume that the utility of dollars is given by the logarithmic dependence hypothesized by Bernoulli. That is,

$$u(\$) = \log_{10}(\$). \qquad (4.11)$$

Figure 4.5 shows the utility values for each possible outcome, and the probability of each event is given along the branch. If the new product is developed, there is a 50–50 chance of success or failure. Based on present

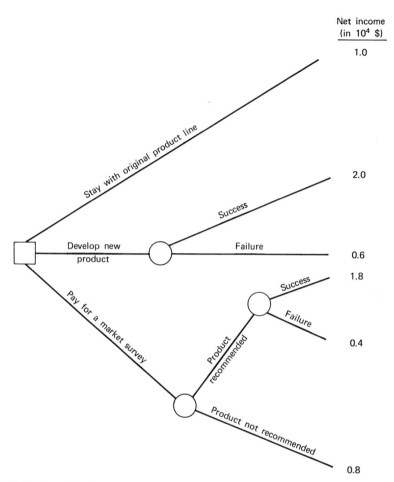

Net income
(in 10⁴ $)

1.0

2.0

0.6

1.8

0.4

0.8

FIGURE 4.4.   A decision tree for new product development. The square □ indicates a point at which a decision must be made, and the circle ○ indicates a point at which chance results occur.

knowledge, therefore, it is equally likely that the market survey will recommend or not recommend the product. However once the survey is performed and there is a recommendation, there is then a 90% chance for success. The expected utility $\langle u \rangle$ for each branch is calculated from the sum $\sum_i p_i u_i$, where p_i is the probability of the ith outcome and u_i is the corresponding utility. It is seen from Fig. 4.5 that the market survey is the highest expected utility branch.

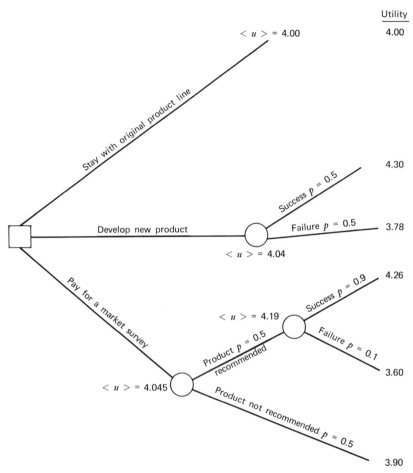

FIGURE 4.5. A decision tree for new product development, showing branch probabilities and expected utilities.

Insurance

Utility theory is used to illustrate the reason for the purchase and sale of insurance. If both the seller and purchaser of insurance were risk indifferent, there would never be a transaction, since an advantage to one is a disadvantage to the other. Most of us are risk averse, and we would prefer to pay more than the odds would warrant to avoid a large loss.

As an example, suppose an individual has $40,000 equity in a house and an additional $2,000 in assets. To simplify the problem we shall assume that during any given year his house remains unmolested or is totally destroyed. From data for houses of similar construction in his area it is found that there is one chance in 100 of destruction during the year. Both the home owner and insurance company are risk averse. The utility function is assumed to be described by the logarithmic dependence given by Eq. (4.11). Figure 4.6 shows the utility scale in the range $5,000–$50,000.

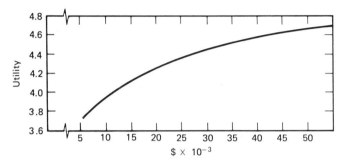

FIGURE 4.6.   The logarithmic utility function would apply to a risk-averse individual.

The utility value for the homeowner's assets at the end of one year without insurance $\langle u \rangle$ is

$$\langle u \rangle = (1 - 10^{-2}) \log_{10}(\$42{,}000)$$

$$+ 10^{-2} \log_{10}(\$2{,}000). \tag{4.12}$$

Equation (4.12) expresses the fact that there is a 0.99 probability of assets of $42,000 and a 0.01 probability of $2,000 at the end of one year. From (4.12) we find that

$$\langle u \rangle = 4.610. \tag{4.13}$$

The dollar equivalent of this utility is the antilog of 4.610 and equals $40,700. If the homeowner were risk indifferent, the mean dollar value to him would be

$$\langle \$ \rangle = (1 - 10^{-2}) \times 42{,}000 + 10^{-2} \times 2{,}000$$

$$= \$41{,}600.$$

We see that the homeowner perceives his assets to be less than a straight-odds calculation because he is reluctant to risk a large loss.

The homeowner is willing to insure his house for \$40,000 if his utility with insurance $\langle u \rangle'$ exceeds $\langle u \rangle$. A maximum annual premium acceptable to the insurer is calculated by equating $\langle u \rangle'$ and $\langle u \rangle$, since a larger premium would reduce $\langle u \rangle'$ below $\langle u \rangle$. If P_{max} is the maximum acceptable premium, we have that

$$\langle u \rangle' = \log_{10}(4.2 \times 10^4 - P_{max}) = \langle u \rangle = 4.610. \tag{4.14}$$

From (4.14), $P_{max} = \$1,300$. If the homeowner was risk indifferent, the maximum premium he would be willing to pay is \$400.

Let us now determine an acceptable minimum premium from the standpoint of the insurance company. The minimum premium is calculated by equating the mean utility for the company when no insurance is offered $\langle u \rangle_{without}$, to the mean utility with insurance, $\langle u \rangle_{with}$. If the company has assets of 10 million dollars, then $\langle u \rangle_{without} = \log_{10} 10^7 = 7.0$. With insurance,

$$\langle u \rangle_{with} = (1 - 10^{-2}) \log_{10}(10^7 + P_{min})$$
$$+ 10^{-2} \log_{10}(10^7 + P_{min} - 4 \times 10^4), \tag{4.15}$$

where P_{min} is the minimum acceptable premium. The assets of the insurance company undergo a small change with or without insurance. Therefore we may assume a linear dependence of utility over the dollar range considered (i.e., the insurance company is risk indifferent for a small change in assets), giving

$$P_{min} \cong (\text{probability of payment}) \times (\text{amount of payment})$$

$$= 0.01 \times 4 \times 10^4 = \$400.$$

We find that any premium between \$400 and \$1,300 is acceptable to both the homeowner and the company.

The strategy choice by the homeowner or insurance company in the previous example (i.e., whether or not to buy or sell insurance) rested with a knowledge of the probability that damage would occur. When the probabilities of various outcomes are known, this is considered decision-making under risk. When the probabilities are not known, this is decision making under uncertainty. In an actual decision making situation, the absence of knowledge about the likelihood of outcomes often results in an acceptance of the status quo. The probable reason that a decision maker is reluctant to take action when the information is incomplete is that a positive step is more susceptible to reproval. However the status quo is not necessarily the best strategy to follow when there is a gap in the knowledge. In the following section we shall consider decision making under uncertainty.

4.3
DECISIONS UNDER UNCERTAINTY

In the insurance problem of Section 4.2 we were able to recommend a course of action because the probabilities of chance outcomes were specified. What do we do when these numbers are not known?

Entropy Maximization

Suppose, for example, there are v voters voting for n candidates for public office, and you are offered a gamble that requires you to specify the preferred candidate for the first voter you encounter. Without any additional information, how would you assign a probability to the likelihood that a given voter prefers a given candidate? We have no basis for believing that any assignment of voters to candidates is more likely than another and, therefore, the most probable distribution of votes would be that distribution with the largest number of ways by which voters can be assigned to candidates. If v_i votes are cast for the ith candidate, then for n candidates the number of ways in which this distribution can occur is given by[4]

$$\text{number of ways} = \frac{v!}{\prod_{i=1}^{n} v_i!}, \tag{4.16}$$

where $\prod_{i=1}^{n} v_i! = (v_1!)(v_2!)(v_3!) \cdots (v_n!)$. With ten voters and two candidates, there is only one way to distribute the voters to a specified candidate, but there are 252 ways to assign the voters when both candidates receive the same number of votes. (As another example, if you had 10 balls to throw randomly into two boxes, the equal distribution would occur 252 times as often as finding all balls in a specified box.) For large v, the distribution of the number of ways to assign voters is sharply peaked about an equal distribution. Therefore the probability p_i that a voter chooses the ith candidate is given by the equal distribution

$$p_i = \frac{1}{n}.$$
(4.17)

All candidates have the same probability of receiving the vote of any given voter and, of course, $\sum_{i=1}^{n} p_i = 1$.

Even without any mathematical calculations it would be reasonable to conclude that $p_i = (1/n)$. If we have no basis for believing that one candidate is preferred to another, then we would logically assign equal values to the probabilities p_i of any individual voting for the ith candidate. Since $\sum_{i=1}^{n} p_i = 1$, this means that $p_i = 1/n$.

If all voters vote for a given candidate, then we can, with certainty, say which candidate any selected voter will choose. This distribution is said to be highly ordered. If the voters are equally distributed, for any selected voter we would then have the lowest chance of guessing which candidate he chose. This distribution is said to be of a low ordering, or highly disordered. The entropy function H given by

$$H = -\sum_{i=1}^{n} p_i \ln p_i$$
(4.18)

is a measure of the disorder of a system. Systems equilibrate in states of maximum entropy, corresponding to high disorder. If H is maximized under the constraint $\sum_{i=1}^{n} p_i = 1$, the result is given by (4.17).

The specification of additional constraints changes p_i. For example, if p_i is the probability that a variable takes on the value x_i, and $\langle x \rangle$ is fixed, the maximization of H gives

$$p_i = \frac{e^{-\lambda x_i}}{\sum_i e^{-\lambda x_i}}$$
(4.19)

where λ is a constant, and p_i is now given by the Boltzmann distribution.[4] Once the probabilities p_i have been determined, then decision making under uncertainty is indistinguishable from decision making under risk.[5]

Seismic Example

As an example of decision making under uncertainty, let us consider the type of building codes we wish to introduce to reduce the destruction that might result from earthquakes. (It is assumed that other parameters, such as soil conditions and distance from a fault, remain constant.) For illustrative purposes, it is assumed that three distinct seismic events may occur over a 20-year planning period in the area under consideration: (a) there may be no quakes of sufficient magnitude to cause building damage, (b) there may be one or more moderate quakes that cause structural damage in the weakest buildings, nonstructural damage (fallen plaster, broken windows, or fallen cornices) in stronger ones, and no damage in the strongest, and (c) strong quakes that cause the weakest buildings to collapse and produce varying degrees of damage in stronger structures. The probabilities of these events are unknown. (In reality, we would be able to obtain an estimate of these probabilities by reviewing earthquake records or consulting with experts.) Our options consist of the introduction of either a strong building code that would add an average of 12% to building costs, a moderate code that would add an average of 6% to building costs, or no code. The stronger the code the less is the likelihood of damage, but the greater is the cost of construction. Table 4.1 enumerates the possible alternatives and events.

Table 4.1
Possible Seismic Events and Alternatives

Alternative	Seismic Event		
	Negligible Shaking	Moderate Shaking	Strong Shaking
No code	Cheapest construction, no damage	Cheapest construction, structural damage	Cheapest construction, total destruction
Moderate code	Moderate cost, no damage	Moderate cost, nonstructural damage	Moderate cost, structural damage
Strong code	Highest cost, no damage	Highest cost, no damage	Highest cost, nonstructural damage

The various combinations of alternative and event are ranked in Table 4.2 in a possible order of preference, with the most preferred at the top of the list. A utility scale from 0 to 10 has been used, with a loss of 1 unit (sometimes referred to as a *utile*) for nonstructural damage, a loss of 2 units for each step in cost; a loss of 6 units for structural damage, and a loss of 10 units for total destruction. (The utility values do not necessarily represent anyone's preferences, but rather are used for illustrative purposes.) Table 4.3 summarizes the utility assignments.

Table 4.2
Utility Ratings for Seismic Events

Alternative and Event	Utility
Cheapest construction, no damage	10
Moderate cost, no damage	8
Moderate cost, nonstructural damage	7
Highest cost, no damage	6
Highest cost, nonstructural damage	5
Cheapest construction, structural damage	4
Moderate cost, structural damage	2
Cheapest cost, total destruction	0

Table 4.3
Utility Table

	Seismic Event		
Alternative	Negligible Shaking	Moderate Shaking	Strong Shaking
No Code	10	4	0
Moderate	8	7	2
Strong	6	6	5

Since the probabilities are unknown, on the basis of our previous discussion we use a probability of 1/3 for each seismic event. Thus the expected utilities for each alternative are

$$\text{No code: } \langle u \rangle = \frac{1}{3}(10) + \frac{1}{3}(4) + \frac{1}{3}(0) = 4.67$$

$$\text{Moderate code: } \langle u \rangle = \frac{1}{3}(8) + \frac{1}{3}(7) + \frac{1}{3}(2) = 5.67$$

$$\text{Strong code: } \langle u \rangle = \frac{1}{3}(6) + \frac{1}{3}(6) + \frac{1}{3}(5) = 5.67$$

From the standpoint of maximization of $\langle u \rangle$, both the moderate and strong codes are equally satisfactory.

How much should be invested for the gathering of additional information? The best possible consequence of having more information is to know precisely which event will occur. This would be perfect information, and the value to be gained in this case would set an upper limit on the amount to be invested.

What is the value of perfect information? If we knew that there would be negligible shaking, we would require no code (no code gives the highest utility when there is negligible shaking), with a corresponding utility of 10; if we knew that there would be moderate shaking we would require a moderate code, with a utility of 7; and if we knew that there would be strong shaking we would require a strong code with a utility of 5. Each seismic event has an a priori probability of 1/3 (based on entropy maximization) and therefore the expected utility of perfect information is:

$$\langle u \rangle = \frac{1}{3}(10) + \frac{1}{3}(7) + \frac{1}{3}(5) = 7.33$$

If we wish to determine the amount of dollars to invest for additional information, we must relate the utility assignments in Table 4.3 to dollars. For example, suppose that in a given community a 6% increase in construction costs corresponds to 2×10^8. Two utiles were associated with a 6% increase and therefore our monetary equivalent for utility is 10^8 per utile. (We are assuming risk indifference over the range of dollars involved in the problem.) The difference in utility between perfect information and the absence of information is $7.33 - 5.67 = 1.66$ utiles, with a dollar equivalent of 1.66×10^8. Therefore, the upper limit to our investment for further information is 1.66×10^8, for this is the value of perfect information.

Other Strategies

Alternatives may be selected on a basis other than the maximization of expected utility, under either risk or uncertainty. If a decision maker tends to be a pessimist (i.e., one who believes that the worst will happen), he may choose a *maximin* approach, wherein he selects the alternative that has the maximum value for its smallest utility. With this choice, under the least desirable condition the utility is greater than for any other alternative. From Table 4.3 we have

Alternative	Minimum Utility (least favorable event occurs)
No Code	0
Moderate	2
Strong	5

The maximin strategy recommends a strong code, for then utility never falls below 5, even under the most adverse conditions.

An optimist would choose the *maximax* approach, which selects the alternative that maximizes the maximum utility. The optimist believes that the best will happen. From Table 4.3

Alternative	Maximum Utility (most favorable event occurs)
No Code	10
Moderate	8
Strong	6

Therefore the optimist would choose the no code alternative.

We may base a decision on how sorry we are that we selected the wrong alternative for a given outcome. *Regret* is defined as the amount that the utility, for any combination of decision and event, differs from the utility for the best decision for that event. If the event is "negligible shaking," the best decision is no code (zero regret), the next best decision is a moderate code (a regret of 2), and the poorest decision is a strong code (a regret of 4). The regret factor is tabulated in Table 4.4.

One way to make a decision is to choose the alternative with the lowest value for the maximum regret. The maximum regret for no code is 5, the

Table 4.4
Regrets

Alternative	Seismic Event		
	Negligible Shaking	Moderate Shaking	Strong Shaking
No Code	0	3	5
Moderate	2	0	3
Strong	4	1	0

maximum regret for a moderate code is 3, and the maximum regret for a strong code is 4. Therefore, we would choose a moderate code. With this alternative, no matter what occurs our regret will not exceed 3.

In this section, various decision strategies were discussed. Mean utility may be optimized or, alternately, a pessimistic, optimistic, or regrettist approach may be used. There is no best strategy; rather, the choice depends on the inclinations of the interested parties.

4.4
INDIFFERENCE RELATIONSHIPS BETWEEN VARIABLES

In the remaining sections of this chapter, various decision problems dealing with incommensurate attributes, that is, attributes measured in different unit systems, will be considered. Several examples and procedures will be described. In Section 1.5 it was stated that indifference relationships may be used to combine attributes into a single objective function, and this approach will be presented in this section. An indifference curve (i.e., line of constant utility) between two variables such as cost and time, may be used to obtain a functional relationship that describes the cost equivalence of time or vice versa. These two variables can then be transformed into a common set of units.

An indifference curve may be determined by establishing equally acceptable conditions. For example, suppose it is found that a trip that costs $30 and takes 2 hours is equally acceptable as a trip that costs $20 and takes 3 hours. If it is assumed that over this range of parameters a linear approximation is reasonable, then only two points are required to establish the constant utility line. From the data we have that

$$c + 10t = 50, \tag{4.20}$$

is an indifference line in the cost–time plane, where $c = $ cost and $t = $ time.

The linear assumption may not be appropriate, and in general, the indifference relationship is given as $g(c)$, where $g(c)$ is the time equivalent of cost (i.e., $g(c)$ is the amount of travel time that may be substituted for a given cost to maintain a constant level of satisfaction). Alternatively, one may use the function $g^{-1}(t)$ as the cost equivalent of time, where g^{-1} is the inverse function of g. A total cost C can be defined as

$$C = c + g^{-1}(t), \tag{4.21}$$

which includes both the actual cost and the cost equivalent of time. Similarly, total travel time T is

$$T = t + g(c). \tag{4.22}$$

In the cost–time plane an indifference line is given by

$$c + g^{-1}(t) = \text{constant}$$

or

$$t + g(c) = \text{constant}$$

since constant total cost or total time corresponds to a fixed value of utility.

For Eq. (4.20) we have that the cost equivalence of time is $10t$. That is, an additional hour of travel time is perceived as an additional cost of \$10. Similarly, the time equivalent of cost is $c/10$. Therefore the total cost C is given by $C = c + 10t$, and the total time T is $T = t + c/10$.

Nonlinear indifference lines might appear as shown in Fig. 4.7, where the direction of maximum increase of utility is orthogonal to the indifference lines.

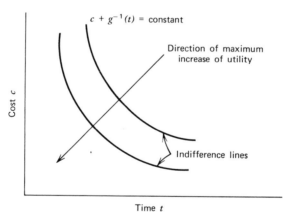

FIGURE 4.7.   Indifference lines showing the direction of maximum increase of utility.

An Example

Once the cost equivalent of time has been defined (or vice versa), a utility function $U(C)$ may be determined for total cost in the manner described in Section 4.2 for a single variable. The object of the decision analysis is to optimize $U(C)$ or $\langle U(C) \rangle$, if risk is involved.

As an illustration, consider the situation where both the cost and time expended on a public project are functions of a single variable N, the number of employees. Time is reduced by having more employees, and cost is increased. Let us assume that the dependencies are

$$t = t_0 + \frac{t_1}{N} \tag{4.23}$$

$$c = c_0 + c_1 N, \tag{4.24}$$

where t_0, t_1, c_0, and c_1 are positive constants. The problem is to determine the value of N that maximizes utility.

To proceed further, it is necessary to define a cost equivalent of time (or vice versa). If a linear equivalence is assumed, such that t hours has an associated cost of (t/R) dollars, where R is constant, then total cost C is

$$C = c + \frac{t}{R} . \tag{4.25}$$

The indifference lines (i.e., constant utility lines) in the cost–time plane are lines of constant C, with slope equal to $(-1/R)$. Constant utility lines are drawn in Fig. 4.8. Utility increases as C decreases, and the direction for increasing utility is given by the arrow in this figure.

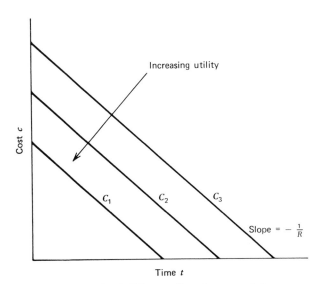

FIGURE 4.8.   Indifference lines for cost and time.

If N is eliminated from Eqs. (4.23) and (4.24), we obtain an additional relationship between c and t

$$c = c_0 + \frac{c_1 t_1}{t - t_0}, \qquad (4.26)$$

which is plotted in Fig. 4.9 as a dashed line. The tangency point A, shown in this figure corresponds to the maximum utility that can be achieved under the constraint imposed by Eq. (4.26). At the tangency point the derivative of the dashed line equals the derivative of the constant utility line. From Eq. (4.26)

$$\frac{dc}{dt} = -\frac{c_1 t_1}{(t - t_0)^2} = \text{derivative for the dashed line}, \qquad (4.27)$$

and for the constant utility line

$$\frac{dc}{dt} = -\frac{1}{R}. \qquad (4.28)$$

Equating (4.27) and (4.28) and solving for t, we find that the time corresponding to maximum utility is

$$t = t_0 + \sqrt{c_1 t_1 R}. \qquad (4.29)$$

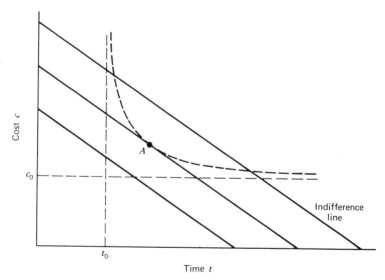

FIGURE 4.9. The constraint expressed by (4.26) is superimposed on the indifference curves.

Substituting this value for t into Eq. (4.23), the optimum number of employees is found to be

$$N = \sqrt{\frac{t_1}{c_1 R}} \qquad (4.30)$$

There are other methods for obtaining this result [see Problem 4.1, or by direct substitution of Eqs. (4.23) and (4.24) in (4.25) and differentiating], but the graphical approach presented in Fig. 4.9 gives a clear understanding of the role of utility in the decision process.

Guaranteed Income

Indifference relationships are used by economists to aid in the prediction of individual or group behavior. For example, a worker may indicate his indifference between income and the number of hours worked as shown in Fig. 4.10.[6] Curves C_1, C_2, and C_3 are indifference lines, that is, the worker is equally satisfied with any combination of work hours and income on a given line. As the number of hours worked increases, a higher income is required to maintain a constant level of satisfaction. The dashed line AO represents the wage rate, where the slope of this line is the hourly salary that the worker is able to earn. If the worker seeks to maximize his utility and there are no other constraints on him, then the point of tangency T gives his income and hours worked.

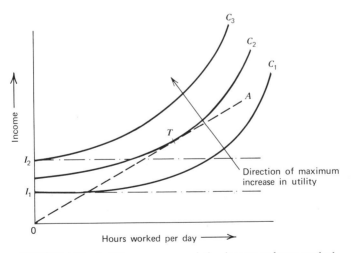

FIGURE 4.10. Indifference curves relating income to hours worked.

Suppose a guaranteed minimum income is introduced, so that if the earned income is below the minimum the worker is paid the difference, and if his income is above the minimum he receives no payment. If the minimum is introduced at income level I_1, and the wage rate remains unaltered, the worker prefers to continue working at point T on curve C_2, since this is a higher utility curve than C_1. If the minimum wage is set at I_2, he prefers to stop working and accept the minimum, because by so doing he moves to curve C_3. He is able to achieve a higher utility by not working, even though his new income is below his earned income.

Consumption and Pollution

As an additional example, consider the selection of a course of action for meeting consumer demands for goods without serious deterioration of environmental quality.[7] A conflict exists because, for given processing techniques and composition of the Gross National Product (GNP), the more we consume the more we pollute. For example, specified air-quality standards may be met by a curtailment in the utilization of automobiles. Is this a good tradeoff? In this section we shall derive expressions for consumption, as a function of time, that reflect both the wish to consume and to reduce pollution.

There are two parts to this problem: (a) a physical relationship between consumption and pollution and (b) the subjective attitudes toward these variables. For the physical equation it is assumed that the rate of generation of pollution is proportional to consumption, and that there is a cleanup rate proportional to the level of pollution. Equation (4.31) follows directly from these assumptions:

$$\frac{dP}{dt} = \dot{P} = \alpha C - \delta P, \qquad (4.31)$$

where P is the pollutant mass, C is the consumption in mass per unit time, t is time, and α and δ are positive constants. The coefficient α is dependent on the technological and processing efficiency. For example, additional recycling would reduce the value for α. It is assumed that technology is constant over the planning period. Coefficient δ may result from either a natural decay process or an imposed cleanup program.

Equation (4.31) is only one feasible technical relationship. Other equations are possible, such as the addition of a constant term to the right-hand side of (4.31) that represents a fixed rate of pollutant removal.

To describe attitudes toward consumption and pollution, it is assumed that the utility U derived over a planning interval from $t=0$ to $t=T$ is given by

$$U = \int_0^T u(C,P)e^{-\gamma t}\,dt, \tag{4.32}$$

where $u(C,P)e^{-\gamma t}\,dt$ is the incremental contribution to utility from t to $(t+dt)$. The factor $e^{-\gamma t}$ represents a discounting of the future relative to the present. If, for example, $\gamma = .05/\text{year}$, this means that the concerns for consumption and pollution 20 years hence are e^{-1} of the present concerns (i.e., we tend to emphasize immediate problems more than future ones). It is possible that some day we may wish to stress the future relative to the present, in which case γ becomes negative.

Our purpose is to determine consumption as a function of time $C(t)$ that optimizes the objective function U subject to the technical relationship expressed by (4.31). It is assumed that the initial pollution level $P(0)$ is given, and that the pollution level at the termination of the planning period $P(T)$ is specified. Neither resource depletion nor population changes are considered, although it is not conceptually difficult to add these modifications to the problem.

In a previous example dealing with the minimization of cost in the presence of a constraint a graphical procedure was presented (see Fig. 4.9). For the consumption–pollution problem, the method of Lagrange multiplier given in Appendix B will be used. From Eq. (B.10), the maximization of (4.32) in the presence of (4.31) is equivalent to the maximization of the integral $\int_0^T L(t,C,P,\dot{P})\,dt$, where

$$L(t,C,P,\dot{P}) = u(C,P)e^{-\gamma t} - \lambda(t)[\dot{P} - \alpha C + \delta P]. \tag{4.33}$$

This maximization leads to the differential equations, (B.12), which are

$$\frac{\partial L}{\partial C} = 0 \qquad \frac{\partial L}{\partial P} - \frac{d}{dt}\frac{\partial L}{\partial \dot{P}} = 0. \tag{4.34}$$

The substitution of (4.33) into (4.34) gives

$$\frac{\partial u}{\partial C}e^{-\gamma t} + \alpha\lambda = 0 \tag{4.35}$$

$$\frac{\partial u}{\partial P}e^{-\gamma t} - \delta\lambda + \dot{\lambda} = 0, \tag{4.36}$$

which, upon elimination of λ, yields

$$\left(\frac{\partial \dot{u}}{\partial C}\right) - (\gamma + \delta)\frac{\partial u}{\partial C} - \alpha\frac{\partial u}{\partial P} = 0. \tag{4.37}$$

The simultaneous solution of (4.31) and (4.37) gives the optimizing $C(t)$ and the corresponding $P(t)$.

Steady-State Solution

The steady-state optimizing solution to (4.31) and (4.37) is obtained by equating to zero the terms differentiated with respect to time. This gives

$$C = \frac{\delta}{\alpha}P \tag{4.38}$$

$$(\gamma + \delta)\frac{\partial u}{\partial C} = -\alpha\frac{\partial u}{\partial P}. \tag{4.39}$$

An advantage to the steady-state solution is that all time periods have the same utility. In other words, we neither sacrifice the future for the present nor the present for the future. It should be noted that even with a steady-state condition, individuals may still have a time preference. The presence of the discount factor γ in Eq. (4.39) means that the time preference affects the optimum steady-state solution.

Equation (4.39) relates the rate of change of utility with respect to consumption $(\partial u/\partial C)$ to the rate of change of utility with respect to pollution $(\partial u/\partial P)$. If utility is expressed in dollars, then $(\partial u/\partial C)$ is the number of dollars per unit of consumption (i.e., the *price* of consumption, where price is defined as cost per unit). Similarly, $\partial u/\partial P$ is the cost per unit time that is associated with pollution. To achieve an optimizing steady-state condition, we must have the ratio of consumption price to pollution cost as specified by (4.39). If $u(C,P)$ is defined explicitly in terms of C and P, the simultaneous solution of (4.38) and (4.39) yields the steady-state values for these variables.

The Utility Function

To proceed further with this problem it is necessary to define a specific $u(C,P)$. A reasonable description of attitudes would include the following conditions: (a) marginal utility per unit time with respect to C, $\partial u/\partial C$, is positive, with decreasing value as C increases and (b) marginal utility per unit time with respect to P is negative with larger magnitude as P increases (i.e., as pollution increases toward a critical value, the concern increases at a faster rate). If we arbitrarily set the range on $u(C,P)$ to be from zero to

1, then zero utility would correspond to $C=0$ or P reaching a critical value P_0. A unity value for utility would occur when $P=0$ and C is large.

Based on these considerations, a reasonable form for $u(C,P)$ is:

$$u(C,P)=\left[1-e^{-C/C_0}\right]\left[1-\left(\frac{P}{P_0}\right)^n\right], \tag{4.40}$$

where $C>0$, $P_0>P>0$, $t>0$, and C_0, P_0 and n are positive constants. Figure 4.11 illustrates the dependencies of u on C and P. The parameter n determines how rapidly concern develops with increasing P. If n is small, then u decreases significantly for small P, whereas for large n, u diminishes only as the critical value P_0 is approached.

If we set u equal to a constant, we obtain the indifference curves for consumption and pollution from (4.40). Figure 4.12 shows these indifference lines. Pollution concern becomes manifest as P_0 is approached,

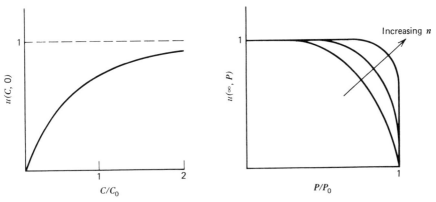

FIGURE 4.11.   Utility dependence on the variables C and P.

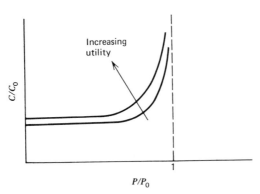

FIGURE 4.12.   Indifference curves for consumption and pollution with a fixed value for n.

whereupon large increases in consumption are required for small pollution increases, if a constant level of utility is to be maintained.

Low Pollution Solutions

A closed-form solution may be obtained when the pollution is low. In this case, $u(C, P)$ is essentially independent of P and (4.37) reduces to

$$\left(\frac{\partial \dot{u}}{\partial C}\right) - (\gamma + \delta)\frac{\partial u}{\partial C} = 0, \tag{4.41}$$

where $u = 1 - e^{-C/C_0}$. The simultaneous solution of (4.31) and (4.41) gives

$$\frac{C(t)}{C_0} = \frac{C(0)}{C_0} - (\gamma + \delta)t. \tag{4.42}$$

The optimizing consumption function is maximum for $t = 0$ and decreases linearly with time. This is true even when there is no discount for the future, that is, when $\gamma = 0$.

If the planning period is long compared to the cleanup time, that is, $\delta T \gg 1$, $C(0)$ is a relatively simple expression and (4.42) becomes

$$C(t) = \left[\frac{\delta}{\alpha}P(T) + (\gamma + \delta)TC_0\right] - (\gamma + \delta)C_0 t. \tag{4.43}$$

The corresponding pollution function is

$$P(t) = \left[P(T) + \alpha C_0 T\right] - \alpha C_0 t + \left[P(0) - P(T) - \alpha C_0 T\right]e^{-\delta t}. \tag{4.44}$$

Figure 4.13 shows how the optimizing $C(t)$ varies according to (4.43) as γ and δ are changed. As the discount factor γ increases (see Fig. 4.13a), the initial value of $C(t)$ increases relative to the final value. This is reasonable, for larger γ means that more significance is given to the present relative to the future. From Fig. 4.13b it is seen that consumption also increases as the cleanup rate δ becomes larger.

Figure 4.14 shows the optimizing pollution function curve derived from (4.44), which applies when pollution is low and $\delta T \gg 1$. This curve was obtained for the boundary condition $P(0) = 0$. $P(t)$ does not depend on γ and increases for increasing cleanup rate δ. The latter result is somewhat surprising in that the optimizing pollution function is greater for higher cleanup rates. This occurs because, as shown in Fig. 4.13b, $C(t)$ increases with δ, thereby increasing $P(t)$.

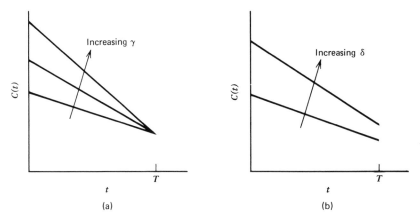

FIGURE 4.13. The optimizing consumption function $C(t)$ as a function of t for various values of the parameters γ and δ, for $\delta T \gg 1$. These curves apply when $P \ll P_0$.

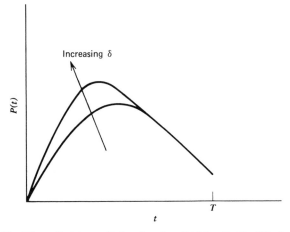

FIGURE 4.14. The optimizing pollution function $P(t)$ for $P \ll P_0$, $\delta T \gg 1$, and $P(0) = 0$.

Computer Solutions

Solutions were obtained for $C(t)$, $P(t)$, and U for the following parameter values:

$\gamma = .02$ years^{-1}, corresponding to a discount factor e^{-1} for 50 years hence;

$n = 5$; $T = 50$ years $=$ planning period

$$\frac{\alpha C_0}{P_0} = .02 \text{ years}^{-1}, \text{ corresponding to approximately 2\% increment}$$
in pollution per year due to consumption;

$$\frac{P(0)}{P_0} = .8$$

Three cases have been considered:

1. The optimum steady-state condition for $\delta = .01$ years^{-1}, corresponding to a pollution decay to e^{-1} over a 100-year period. This results in a consumption C equal to $.4 C_0$.
2. The optimizing solutions for $P(T) = P(0)$ and $\delta = .016$ years^{-1}, corresponding to a pollution decay to e^{-1} over approximately 60 years.
3. An imposed consumption growth that doubles consumption in 20 years, for $\delta = .01$ years^{-1} and $C(0) = .4 C_0$.

The results are shown in Figs. 4.15 and 4.16. For case 1 both consumption and pollution are constant. For case 2, consumption grows monotonically and pollution dips slightly and then returns to its initial value at the end of the planning period. For case 3, consumption has an imposed growth and pollution increases to the critical value P_0 in slightly over 30 years. The utilities integrated over 50 years are, for each case, equal to

$$U_{(1)} = 8.5 \qquad U_{(2)} = 12.3 \qquad U_{(3)} = 6.5.$$

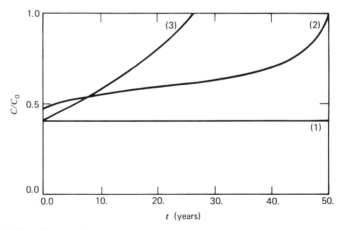

FIGURE 4.15. Consumption as a function of time for the parameter values $\gamma = .02$ years^{-1}, $n = 5$, $T = 50$ years, $(\alpha C_0 / P_0) = .02$ years^{-1}, and $[P(0)/P_0] = .8$. The three cases are: (1) optimum steady-state for $\delta = .01$ years^{-1}, (2) optimizing solutions for $\delta = .016$ years^{-1}, and (3) imposed growth of consumption, doubling in 20 years, with $\delta = .01$ years^{-1} and $C(0) = .4 C_0$.

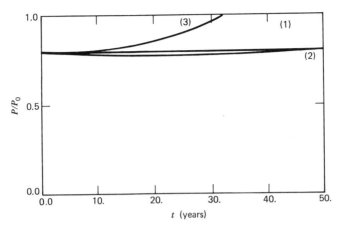

FIGURE 4.16. Pollution as a function of time for the three cases cited in Fig. 4.15.

An exponential growth in consumption yields the smallest utility, and a decrease in the cleanup time from 100 to 60 years provides a marked increase in utility for the optimum solution.

As with all models, this model of the conflict between consumption and pollution is a compromise between reality and simplicity. It is important to recognize the shortcomings of one's analysis. For example, one approximation is the high level of aggregation. There is no distinction between types of pollutants, between items of consumption, or with regard to the distribution of goods among sectors of the consuming public. The specific forms of the functional dependencies chosen for the model [Eqs. (4.31) and (4.40)] represent additional approximations to reality. It has also been assumed that the parameter values, such as those that might be influenced by technological change or by altered individual attitudes, remain constant over the planning interval.

4.5
WEIGHTING UTILITIES

When there are n incommensurate attributes, a single utility expression U is often written as the linear, weighted summation of individual utilities

$$U = \sum_{i=1}^{n} w_i u_i = w_1 u_1 + w_2 u_2 + \cdots + w_n u_n, \tag{4.45}$$

where U is the objective function, u_i is the utility associated with the ith attribute, and w_i is a weighting factor assigned to u_i. In Section 1.5, weights were assigned by a pairwise comparison between attributes, and other methods for weight assignment are discussed in the literature.[8]

Viewpoint Triangle

It may not be possible to arrive at a single set of weights that represent the views of the interested parties. Rather, it might be preferable to indicate the dominant solutions for each of a variety of different weighting schemes. This approach was taken in a Stanford University land-use problem,[9] in which the economic, social, and evironmental factors were represented by the following nine attributes:

Economic Factors
1. Income to Stanford University.
2. Income to local communities.
3. Employment (additional jobs created).

Social Factors
4. Housing (number of units of low-, middle-, and high-income housing).
5. Traffic (number of automobile trips generated).
6. Recreation (quantity and quality of recreational facilities).
7. Social effects (the extent to which different socioeconomic groups are brought together).

Physical Environment Factors
8. Natural effects (biological, geological, and pollution impacts).
9. Esthetics (detractions to the visual quality of the area).

The placement of the nine attributes under the three categories is somewhat arbitrary, and the particular breakdown that was chosen was based on groupings considered reasonable by those involved in the study.

Eleven alternative development plans were analyzed, to represent the range of suggested uses. The plans may be identified by their primary emphasis as follows:

Plan Number	Emphasis
1, 2, 3	Open space and recreational
4, 5, 6	Residential
7, 8, 9, 10	Industrial and commercial
11	Academic reserve

Each attribute was measured for each plan, by using either some physical scale or a worth estimate. These measurements were then converted to utility ratings on a scale of 1–5, with 5 corresponding to the highest utility.

The next step was to assign integral weights to the various attributes to represent the relative importance of each attribute. A normalization condition $\sum_{i=1}^{9} w_i = 100$ was used, where w_i is the weight for the ith attribute. The sum S_j

$$S_j = \sum_{i=1}^{9} w_i u_{ij} \qquad j = 1 \text{ to } 11,$$

was calculated for each of the j plans, where u_{ij} is the utility for the ith attribute under the jth alternative plan. With any given weighting scheme one alternative dominated the others by giving the highest value for S_j. Small changes in the weights did not change which alternative was dominant, and, therefore, similar weighting schemes that led to the same alternative were grouped into a single cluster. Clustering is a technique used in pattern recognition.[10] It was found that 94 clusters were sufficient to represent all of the possible weightings.

To present the results of the study to a decision-making body, the various weighting schemes and dominant alternatives were embodied on a planar figure. This necessitated regrouping the nine attributes under the following three factors: (a) economic, (b) social, and (c) environmental.

Figure 4.17 illustrates a *viewpoint triangle*, representing different weights assigned to economic, social, or environmental impacts. Each vertex of the triangle represents a 100% emphasis on the corresponding factor. For example, the lower left-hand vertex is equivalent to a viewpoint with all of the weight placed on economic factors. As the triangle is traversed from a vertex to the opposite side, the percentage of emphasis on the vertex impact decreases from 100 to 0. The dashed lines within the triangle correspond to an emphasis of 80%, 60%, 40%, and 20%. For example, the highest horizontal line corresponds to an 80% emphasis on social factors.

The numbers displayed on Fig. 4.17 show where the corresponding alternative is dominant on the viewpoint triangle. (Letter T refers to alternative 10.) For example, a viewpoint with 33% emphasis on economic, 25% on social, and 42% on environmental impacts leads to alternative 2 as dominant.

A land-use decision involves a recognition of both the requirements of the land owner and the desires of the residents in the neighboring communities. A survey of 4600 registered voters in contiguous areas included

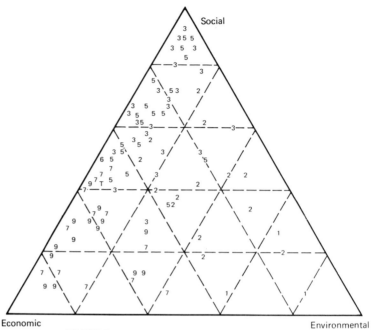

FIGURE 4.17. Land-use study viewpoint triangle.

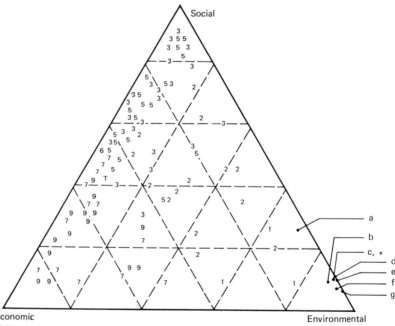

FIGURE 4.18. Community positions located on the viewpoint triangle. *Regional Community viewpoint, a. East Palo Alto, b. Mountain View, c. Los Altos Woodside, d. Portola Valley, e. Ladera, f. Menlo Park, Palo Alto, g. Altherton, Los Altos Hills.

the questions, "What use do you think should be made of the peninsula foothills: (a) residential, (b) park/open space, (c) industrial/commercial, or (d) no opinion." An individual responding (a) would be located at the top of the viewpoint triangle, a (b) response would be at the lower right, and a (c) response would be at the lower left. The viewpoints of the neighboring communities were located on the viewpoint triangle based on the percentage of people in each community that responded (a), (b), or (c) to the question. Figure 4.18 illustrates the locations of community viewpoints. Primary interest is on the *environmental* impact, a small consideration is given to *social* factors, and a negligible interest to *economic* effects. The community with the largest deviation from the average viewpoint is East Palo Alto, primarily a black neighborhood with a lower average income than the other areas. For this example the community preferences lie in the region where alternatives 1 and 2 are dominant.

4.6
SUMMARY

In this chapter a number of different decision problems were presented, along with various methods that might be used to resolve a choice between alternatives. For a single attribute, when risk or uncertainty is involved, an approach is to establish utility values for the possible outcomes and then select a decision strategy that either maximizes expected utility or sets some other constraint on utility (e.g., maximizes minimum utility, minimizes regret, etc.) When incommensurate attributes are considered, a single utility function may be derived from indifference relationships between the attributes. If a constraint exists, defining a relationship between attributes, the problem becomes one of optimizing utility in the presence of the constraint. An approach that may be used for a large number of attributes is to express utility as a linear, weighted sum of the individual utilities.

REFERENCES

1. H. Chernoff and L. E. Moses, *Elementary Decision Theory*, Wiley, New York, 5th print. (1967), Ch. 5.
2. J. von Neumann and O. Morgenstern, *Theory of Games and Economic Behavior*, Princeton U. P. (1953).
3. H. Raiffa, *Decision Analysis*, Addison-Wesley, Menlo Park, Calif. (1968), pp. 10–27.

4. E. Parzen, *Modern Probability Theory and its Applications*, Wiley, New York, 9th print. (1967), pp. 39–40.

5. E. T. Jaynes, "Prior Probabilities," IEEE Trans. on Systems, Science and Cybernetics, *SSC*-4, pp. 227–241 (Sept. 1968).

6. C. E. Ferguson and F. C. Maurice, *Economic Analysis*, Irwin, Georgetown, Ontario (1970), Ch. 4.

7. R. Fleming and R. H. Pantell, "The Conflict Between Consumption and Pollution," IEEE Trans. on Systems, Man and Cybernetics, *SMC*-4, *pp.* 204–208 (1974).

8. Additional discussion of methods for relating one attribute to another is presented in R. L. Ackoff, *Scientific Method for Optimizing Applied Research Decisions*, Wiley, New York (1962), pp. 76–93.

9. M. Neering et al., "Stanford Land Use Study," *Socio-Econ. Plan. Sci.* **6**, 409–419 (1972).

10. M. M. Astraban, "Speech Analysis by Clustering," Computer Science Department, Stanford, Memo AI-124 (May 1970).

PROBLEMS

4.1 Using the method of Lagrange multiplier, derive Eq. (4.30) by minimizing Eq. (4.25) under the constraint given by Eq. (4.26).

4.2 Consider the gambler and conservative illustrated in Fig. 4.3. Each of these individuals has $750 and is offered a "double-or-nothing" bet. What odds would each want in order to accept the bet? Estimate approximate values for utilities from the curves in Fig. 4.3. What odds would each want if they started with $500?

4.3 An X ray of a patient's lung indicates a tumorous growth. The physician may order surgical removal of the tumor or he may do nothing. In the opinion of the physician there is a 90% chance that the tumor is benign and, therefore, no operation is required. If the tumor is cancerous, the patient will die without the operation, and with the removal of the growth he has an 85% chance for recovery. The operation itself involves some risk, in that there is a 5% chance that the patient will not recover from the surgery. Draw the decision tree and determine a course of action based on maximization of utility.

4.4 With reference to Fig. 4.4, if the dollar amounts are in units of 10^6 dollars rather than 10^4 dollars, and the utility function, Eq. (4.11), remains the same, and the probabilities shown in Fig. 4.5 remain the same, what is the recommended course of action for the manufacturer? Does the recommended course change when the units are 10^6 rather than 10^4 dollars?

4.5 For the cost–time problem discussed in Section 4.4, suppose the cost equivalent of time is given as $k \ln(t - t_0)$ for $t > t_0$ and k a positive constant. Draw the indifference curves in the cost–time plane along with the constraint shown by the dashed line in Fig. 4.8. Determine the optimum number of employees from either Eq. (4.23) or (4.24).

4.6 Various site locations have been considered for the development of faculty–staff housing on the Stanford University campus. The accompanying chart indicates ratings that were given to each site with respect to seven different attributes.

Consider only those sites that are not dominated in the ratings by an alternative site. For these sites, only three attributes need be considered. Why? Draw a viewpoint triangle with each of these three attributes at a different vertex, and indicate the regions where the various sites are dominant.

Site Ratings Chart

Site	LAND-USE MASTER PLAN				CIRCULATION MASTER PLAN			
CRITERIA & SITE	Compliance	Availability	Compatibility	Physical Usability	Compliance	Accessibility to Campus Core	Accessibility to Off-Campus Destinations	Total Score
A	1	1	1	1	1	2	1	8
B	1	3	1	1	1	1	1	9
C	1	2	1	1	1	1	2	9
D	1	3	1	2	1	1	1	10
E	2	1	1	2	1	3	1	11
F	3	3	1	1	1	1	2	12
G	3	1	2	2	1	3	1	13
H	3	2	2	2	1	2	1	13
I	3	1	3	2	1	3	1	14
J	2	3	3	2	2	2	2	16
K	2	3	3	2	2	2	3	17
L	2	1	3	3	2	3	3	17
M	3	1	3	2	3	3	3	18

Legend: 1 = good   *Note*:   Low score = good
               2 = fair   High score = poor
               3 = poor

Explanation of Terminology
1. *Compliance* (with master plans) indicates whether or not the proposed medium density housing will require changes in the approved master plans.
2. *Availability* refers to present use of the site—whether it is encumbered with structures, and if so, their expected removal date, or whether the site is vacant.
3. *Compatibility* pertains to the ability of medium density housing to be a good neighbor with a minimum of friction in the area.

4. *Physical usability* has dual importance: (a) whether utilities are readily available and (b) whether the site will need major preparation, such as grading.
5. *Accessibility to campus core* refers to the ease with which residents of a specific site can circulate, preferably on foot, to and from campus destinations of interest to them, such as the post office, bookstore. Faculty Club, libraries, Memorial Auditorium, and Memorial Church.
6. *Accessibility to off-campus destinations* is indicative of the ease with which residents can move on and off the campus, as well as the most direct route possible by which they can be reached by vehicles and off-campus persons such as visitors, deliverymen, and emergency personnel.

4.7 Suppose our utility for dollars is given by

$$u(\$) = 1 - e^{-bx}$$

where $b = 10^{-5}$ dollars^{-1}, and x is in dollars. For the insurance problem discussed in Section 4.2 where you have \$40,000 in equity in a home and \$2000 in additional assets, determine the perceived value of your assets and the insurance premium you would be willing to pay. Assume a .01 probability of losing your house in a given year.

4.8 For Problem 4.7, suppose that the premium is \$2000. Determine whether or not you would pay this premium based on: (a) maximization of expected utility, (b) a maximin approach, (c) a maximax approach, and (d) a regret approach. Assume that the utility function is the same as for Problem 4.7.

Appendix A □ A Case Study

This appendix has been contributed by Dr. Morteza Kashef, who performed his thesis research under my supervision. His thesis topic dealt with the health-care delivery system in San Mateo County, California. As part of his effort he was employed by the County Department of Health and Welfare, and served on a task force that had the responsibility for planning the future of the county's role in health-care delivery. I requested that Dr. Kashef write of his experiences, with particular emphasis on the difficulties encountered by the task force. It is only through direct contact with practitioners and planners that the systems analyst can place his tools in proper perspective. An approach that appears appropriate in academic isolation may be useless as a practical measure.

A.1
INTRODUCTION

It is desirable for planners to understand and be capable of utilizing systems techniques, but there is often a significant gap between the formalism and the application. The purpose of this appendix is to describe some "real world" problems encountered in planning for health-care delivery in San Mateo County, California.

In the latter part of 1972, a task force was appointed by the director of the San Mateo County Department of Public Health and Welfare to study the county's medical care delivery system (MCDS). The objective of the task force was to develop a blueprint for the future of the county government's role in medical care delivery. The need for the study became apparent when the county government realized the necessity for reexamination of its MCDS in the light of many changes that had occurred in

171

the financing and delivery of medical care since the establishment of the existing system. Several important changes were the advent of Medicare and Medicaid and the oversupply of hospital beds in the county.

Various problems evolved during the course of the study, and it is likely that many planning situations would generate similar difficulties. The problem areas included:

1. Dealing with external pressures.
2. Determining the composition of the planning body.
3. Incorporating financial constraints.
4. Data handling.

In the following pages each of these topics is discussed in the context of the plan for the San Mateo County MCDS. First, however, a brief history of the health care delivery system in San Mateo County is presented.

A.2
BACKGROUND INFORMATION

Prior to 1965, local governments (i.e., county and/or city) had the full responsibility for providing medical care to the indigent sick. This responsibility was mandated by state law. To meet this responsibility, county governments constructed elaborate medical care systems for the poor. There were essentially two distinct systems of medical care; one for the poor and the other for the nonindigent. These two systems were segregated, for the nonindigent were not allowed to use the county system, even though there might be a willingness to pay for the services.

The nucleus of the county MCDS was a county hospital where the patient received all of his medical care. County hospitals were generally overcrowded and understaffed. Therefore county hospitals were identified with second-class medical care.

Medicare and Medicaid legislation changed this picture. The former program provided medical benefits for individuals covered by social security, and the latter provided assistance to the economically disadvantaged. There was a dramatic impact on the county MCDS for the reasons:

1. These programs shifted the major responsibility for providing care to poor from the local to the state and federal governments.

2. Under these new programs the poor were free to choose any provider, thereby allowing the poor to use noncounty hospitals and physicians.

Under the new legislation, local governments were required to "buy out" their responsibility to the poor by contributing some of their property tax dollars to the Medicare and Medicaid programs. The primary effect on local governments was a diminished role in the provision of care to the poor. However there was still a small residual needy group of people who were not covered under either program. This included transients who did not meet residence requirements for Medicaid, and individuals incarcerated in local jails.

To discourage local governments from dismantling their MCDS, the State of California created the "option" program. Under this program, the county MCDS would:

1. Operate as a private provider, giving service to all economic sectors.
2. Charge the cost of services to all people receiving care and try to collect such charges.
3. For a fixed annual premium, the state would reimburse the county for all unrecovered costs.

The result of the option program was that San Mateo County opened its system to the nonindigent and made many improvements so as to make the facility attractive to all people.

The poor began to desert the county system quite rapidly, and despite improvement efforts, private patients did not fill the void. The result was a decrease in utilization of the county system and an increase in the private sector. The private sector, expecting a surge of new patients, overexpanded its facilities.

The option plan, whereby the State of California reimbursed the counties for unrecovered costs, was eliminated in 1971. This change had a dramatic impact on the counties' finances, whereupon they tightened their standards for the provision of care to the poor. Some counties closed their public hospitals.

The 1972 task force appointed by the director of the San Mateo County Department of Public Health and Welfare was to suggest possible future courses for the county's role in health-care delivery. Various alternatives that were to be considered included:

1. Whether to continue to maintain a single county hospital.

2. Whether to close the county hospital and establish neighborhood health centers run by the county.

3. Whether to pay private physicians and hospitals from county funds to care for the indigent patients not covered by Medicare or Medicaid.

The criteria for judging the relative merits of each alternative were:

1. Cost to the county.
2. Cost to the patients.
3. Quality of medical care provided.
4. Convenience (i.e., travel time) to the patients and staff.
5. Degree of control over the health care delivery system kept by the County Department of Public Health and Welfare.

A.3
PRACTICAL PROBLEMS OF PLANNING

Dealing with External Pressures

A central issue to the future planning for the MCDS in San Mateo County concerned the fate of the county hospital. This hospital provided medical care for the poor for over half a century, and it symbolized a commitment by the county to the medically indigent. Since the question of the closure of the hospital had been raised in earlier years, there were many people and groups in the county who were very sensitive to this issue. For example, a strong position was taken by the county's major newspaper against closure. The doctors employed by the county hospital formed another vociferous group against closure.

The Department of Public Health and Welfare of the county had the administrative responsibility for any health-care program by the county. The department fully appreciated the political climate and decided to approach the problem cautiously. Creation of the special task force was a direct recognition of the importance of the problem. The relationship of the planning task force to the county heirarchy is displayed in Fig. A-1. The task force was a temporary study group composed of the department's staff. They were responsible to the director of the Health and Welfare Department, who was, in turn, responsible to the county manager and the Board of Supervisors. The Board of Health and Welfare serves as an advisory group to the Health and Welfare Department, as well as to the Board of Supervisors.

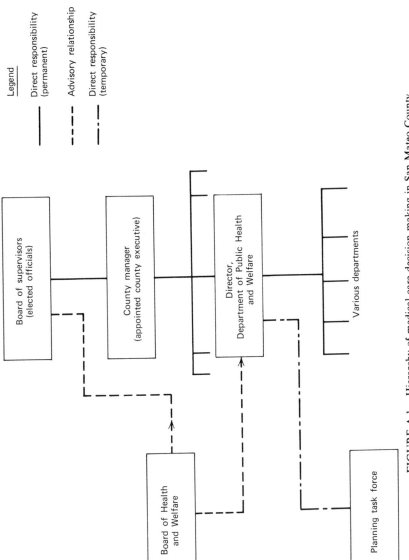

Legend

Direct responsibility
(permanent)

Advisory relationship

Direct responsibility
(temporary)

Board of supervisors
(elected officials)

County manager
(appointed county executive)

Director,
Department of Public Health
and Welfare

Various departments

Board of Health
and Welfare

Planning task force

FIGURE A.1. Hierarchy of medical care decision making in San Mateo County.

175

To insulate the task force from undue lobbying pressures, the following measures were taken:

1. The individuals selected for the task force were not involved in county politics.
2. The task force was made responsible to an administrative officer rather than to a political body.

There was a potential danger, however, of too much isolation from the political process, since it could result in plans that would be politically unacceptable. To counteract this possibility, the task force scheduled periodic information sharing sessions with primary interest groups, such as the county hospital medical staff, patients using the public health facilities, political leaders, and other community organizations. During these sessions the task force shared its findings and opinions as they were developed, allowing and welcoming feedback and criticism. Many ideas were modified and some abandoned as a result of these exchanges. In retrospect, this strategy was very useful and informative.

Determining the Composition of the Planning Body

A decision-making body can spare itself many political problems and harassments by proper selection of the planning body. In dealing with public issues such as health care to the poor, the credentials of the authors of the plan are extremely important. The personalities and the educational backgrounds of the planners are favorite targets of those interest groups that disagree with the planning document. This particular task force included people with backgrounds in medicine and social service, which helped to establish the credibility of the planning group. In addition, a reputable consulting firm was hired to work with the task force, not so much for their expertise, but to instil public confidence in the process.

In most social systems planning there are a large number of diverse interest groups. With regard to the MCDS for San Mateo County, these included:

1. Physicians serving in the county hospital.
2. Other employees of the county hospital.
3. Free clinics serving the county.
4. Other health-care providers.
5. The indigent.
6. Taxpayers.

7. The County Medical Society.
8. The County Comprehensive Health Planning Council.
9. The Board of Supervisors.

On the one hand, it would be desirable to have an individual from each sector involved in the planning so as to incorporate the interests of all parties. Also, it might facilitate acceptance of the plan if all segments felt represented. However this makes for an unwieldy group. Too many diverse viewpoints can lead to interminable bickering. Each expert has his own vocabulary and communication may be difficult. As the number of participants increases, compromise and resolution of issues becomes more difficult.

The planning task force was composed of four people, three of whom (physician, nurse, and social worker) represented interested parties. The fourth member (Dr. Kashef), was presumed neutral. The task force maintained an exchange of ideas with other parties by means of information sharing sessions, as described previously. This group worked extremely well together.

Incorporating Financial Constraints

The task force was informed that the county planned to budget $4,000,000/annum for the MCDS; however, it was suggested that as much as $6,000,000 would be acceptable if the additional cost were properly justified. It was difficult to include any consideration of financial constraints in the planning. Several members of the task force believed that it involved a compromise of principles (i.e., it was a matter of personal integrity) if planning were altered for the sake of monetary limitations. The plan adopted by the task force was approved before cost estimates became available. It was felt that it was the responsibility of the Board of Supervisors to determine what could or could not be afforded. In the plan, concern was expressed that one or another course of action might be expensive, but no attempt was made to limit expenditure to the amount specified by the county budget.

Data Handling

The evaluation of alternative plans required information that was often unavailable or inadequate. For example, it would have been useful to know the numbers and distribution of prospective consumers and the dependence of the utilization of health-care facilities on travel time and cost to the patient. Not only was information necessary to describe existing needs, but projections had to be made for future health-care facilities.

Most of the data in the health-care field has been gathered primarily for clinical rather than management purposes. It is generally difficult to adapt data from another source to one's own requirements. There is a lack of uniformity in the definition of terms, and different sources are often contradictory. For example, the extent to which economic status influences utilization of health-care services is not well documented.

It was expensive and time consuming for the task force to gather information. In some instances, expert opinion was used to bridge gaps in the data. The task force became particularly knowledgeable concerning the past history of the utilization of county services by different segments of the population. This effort served a useful purpose, for the task force was well prepared to respond to questions in this area.

A.4
SUMMARY AND CONCLUSIONS

The purpose of this appendix has been to discuss practical difficulties encountered in social systems planning. Some insights can be gained only by working with practitioners and planners. For the academician, field contact will improve his ability to communicate his ideas, will teach him to compromise his procedures so as to make them more palatable to others, and will heighten his sensitivity to the desires of the interested parties. A familiarity with economics, decision theory, and conservation equations will have little consequence unless it is combined with these other ingredients.

PROBLEMS

A.1 Suppose a planning group, of which you are a member, produces a document that is opposed by the local newspaper. Consider the pros and cons of the following courses of action:

(a) Ignoring the newspaper.
(b) Writing letters to the editor, presenting your viewpoint.
(c) Placing advertisements in the paper, highlighting your conclusions and reasons for them.

(d) Visiting the editor of the paper, and attempting to persuade him to modify his position.

(e) Organizing a boycott of the paper.

A.2 If you are a chairing a planning group, and you feel that one member of this group is constantly drifting on to irrelevant issues, what methods might you use to correct this problem? Several suggestions are:

(a) Tell the member that his discussion is irrelevant, and move on to the desired topic.

(b) Announce at the outset of the discussion that, due to time limitations, each participant will be limited to a specified time period.

(c) Forget to send meeting notices to the troublesome member.

(d) Wait patiently for the person to finish, and then move on to the desired topic.

(e) Arrange for another member of the group to point out that the discussion is rambling.

(f) Tolerate the irrelevant discussions.

Discuss these alternatives.

A.3 As a systems analyst you are attempting to determine attitudes towards a particular condition from members of a planning group. You request an ordering of alternatives with an assigned numeric value. Several individuals object on the basis that attitudes are not quantifiable. Would you prefer to:

(a) Drop the whole idea?

(b) Tell the objecting members that you are unable to proceed without their cooperation?

(c) Try to explain why you want to quantify attitudes?

(d) Obtain information from those individuals who are willing to respond, and ignore the others?

Why would you choose one course of action in preference to another?

A.4 Upon presenting your plan for health-care delivery to a public meeting, a member of the audience claims that the plan is invalid because there was neither a black nor Chicano representative on the planning group. How would you respond to this statement, and why?

Appendix B □ Optimization With Constraints (Method of Lagrange Multiplier)

First, we shall consider the maximization (or minimization) of $f(x,y,z)$ given by the constraint $g(x,y,z)=0$. Without the constraint

$$df = \frac{\partial f}{\partial x} dx + \frac{\partial f}{\partial y} dy + \frac{\partial f}{\partial z} dz = 0 \quad \text{(at the maximum)}. \qquad (B.1)$$

For x, y, and z independent variables (unconstrained), dx, dy, and dz may take on independent values so that (B.1) implies

$$\frac{\partial f}{\partial x} = \frac{\partial f}{\partial y} = \frac{\partial f}{\partial z} = 0 \quad \text{(at the maximum)}.$$

If $g(x,y,z)=0$ is a constraint, then for all values of the variables:

$$dg = \frac{\partial g}{\partial x} dx + \frac{\partial g}{\partial y} dy + \frac{\partial g}{\partial z} dz = 0. \qquad (B.2)$$

From (B.2) we see that dx, dy, and dz are no longer independent. If we solve (B.2) for dz, we have

$$dz = - \frac{\dfrac{\partial g}{\partial x} dx + \dfrac{\partial g}{\partial y} dy}{\dfrac{\partial g}{\partial z}}. \qquad (B.3)$$

The substitution of (B.3) into (B.1) gives

$$\left[\frac{\partial f}{\partial x} - \frac{\dfrac{\partial f}{\partial z} \dfrac{\partial g}{\partial x}}{\dfrac{\partial g}{\partial z}} \right] dx + \left[\frac{\partial f}{\partial y} - \frac{\dfrac{\partial f}{\partial z} \dfrac{\partial g}{\partial y}}{\dfrac{\partial g}{\partial z}} \right] dy = 0. \tag{B.4}$$

In (B.4) we may consider dx and dy to be independent, since all values of dx and dy will satisfy the constraint (B.2). Therefore from (B.4) we require that

$$\frac{\partial f}{\partial x} - \frac{\dfrac{\partial f}{\partial z}}{\dfrac{\partial g}{\partial z}} \left(\frac{\partial g}{\partial x} \right) = 0 = \frac{\partial f}{\partial y} - \frac{\dfrac{\partial f}{\partial z}}{\dfrac{\partial g}{\partial z}} \left(\frac{\partial g}{\partial y} \right). \tag{B.5}$$

Let $\lambda \equiv (\partial f / \partial z)/(\partial g / \partial z)$ evaluated at the maximum ($\lambda =$ the Lagrange multiplier). Therefore from (B.4) and (B.5) we may write

$$\frac{\partial f}{\partial x} - \lambda \frac{\partial g}{\partial x} = \frac{\partial f}{\partial y} - \lambda \frac{\partial g}{\partial y} = \frac{\partial f}{\partial z} - \lambda \frac{\partial g}{\partial z} = 0. \tag{B.6}$$

Equation (B.6) is equivalent to the maximization of

$$f(x,y,z) - \lambda g(x,y,z). \tag{B.7}$$

The simultaneous solution of (B.7) and $g(x,y,z) = 0$ maximizes $f(x,y,z)$ in the presence of the constraint.

From (B.6) we see that if the constraint function g is allowed to change, then

$$\lambda = \frac{df}{dg} \quad \text{(evaluated at the maximum)}.$$

When the function to be maximized f is a net benefit, then the constant λ is the marginal change in net benefit for a marginal change in the constraint function, evaluated at the point where f is maximized. With f measured in dollars, λ has the dimensions of price, that is, dollar change in net benefit per unit change in constraint, and is known as the *shadow price*.

For example, if the constraint imposes a limitation on the amount of floor space available for a project, then the shadow price measures the change in net benefit with a unit change in the available floor space, at the point where benefit is a maximum. Shadow prices are used to suggest which constraints should be modified. If the shadow price for a floor-space constraint is much higher than the shadow price for a manpower constraint, then a unit of additional floor space will produce a larger increment in net benefit than an additional unit of manpower.

 If the function to be maximized and the constraint are expressed as integrals between constant limits, that is,

$$\text{maximize:} \quad \left. \begin{array}{l} \int_a^b f(x,y,z,y',z')\,dx \\[2ex] \text{given:} \quad \int_a^b g(x,y,z,y',z')\,dx = 0 \end{array} \right\}, \qquad \text{(B.8)}$$

where y and z are functions of x and $y' = (dy/dx)$, $z' = (dz/dx)$, then by a method analogous to the derivation of (B.7) it may be shown that (B.8) is equivalent to

$$\text{maximize:} \quad \int_a^b L(x,y,z,y',z')\,dx, \qquad \text{(B.9)}$$

where

$$L(x,y,y',z,z') = f(x,y,z,y',z') - \lambda g(x,y,z,y',z'). \qquad \text{(B.10)}$$

If an integral is to be maximized and the constraint is not an integral, that is,

$$\text{maximize:} \quad \left. \begin{array}{l} \int_a^b f(x,y,z,y',z')\,dx \\[2ex] \text{given:} \quad g(x,y,z) = 0 \end{array} \right\} \qquad \text{(B.11)}$$

Equations (B.9) and (B.10) still apply, except that λ now becomes $\lambda(x)$, a function of x.

 Maximization of the integral given by Eq. (B.9) leads to the differential equations

$$\left.\begin{array}{l} \dfrac{\partial L}{\partial y} - \dfrac{d}{dx}\dfrac{\partial L}{\partial y'} = 0 \\[2mm] \dfrac{\partial L}{\partial z} - \dfrac{d}{dx}\dfrac{\partial L}{\partial z'} = 0 \end{array}\right\} \qquad \text{(B.12)}$$

The solution to (B.12) gives the functions $y(x)$ and $z(x)$ that maximize the integral in the presence of the constraint, as expressed by (B.8) or (B.11).

□ Index